Y0-BDG-048

Tweeting DaVinci

Published in the United States by:

Da Vinci Press

New York, New York 10011

www.davincipress.org

email: davincipress @gmail.com

email: tweetingdavinci@gmail.com

Twitter: @tweetingdavinci

www.tweetingdavinci.com

Printed in South Korea

Book design and layout: Donald Partyka

Cover design and illustrations: Francesco Filippini

ISBN 978-1-940613-00-0

Library of Congress Control Number: 2014938414

1. Italy-tourism. 2. Italy-geology. 3. Geology-Italy. 4. Classical Studies-Italy. 5. Etruscans-Italy. 6. Dante-Divine Comedy. 7. Gems-Gemology. 8.Virgil-Aeneid. 9.Leonardo da Vinci-Virgin of the Rocks. 10. Boltraffio, d'Oggiono-art history. 11. Grottoes-geology. 11. Lightning-Earthquakes. 12. Geomagnetic fields-Magnetite.

Tweeting Da Vinci

Ann C. Pizzorusso

Dedicated to
The Dream Team
Leonardo Da Vinci
The Etruscan priests
Virgil
Dante

and
my mother
Ada Di Loreto
who brought me to collect spring water

CONTENTS

FOREWARD.
page IV

ACKNOWLEDGEMENTS.
page VII

INTRODUCTION
ITALY: A LAND BORN OF THE SEA
page 1

I.
A LANDSCAPE DIVINE:
ETRUSCANS AND THEIR ENVIRONMENT
page 11

II.
HELL ON EARTH:
INTO VIRGIL'S UNDERWORLD
page 59

III.
PARADISE BEJEWELED
page 91

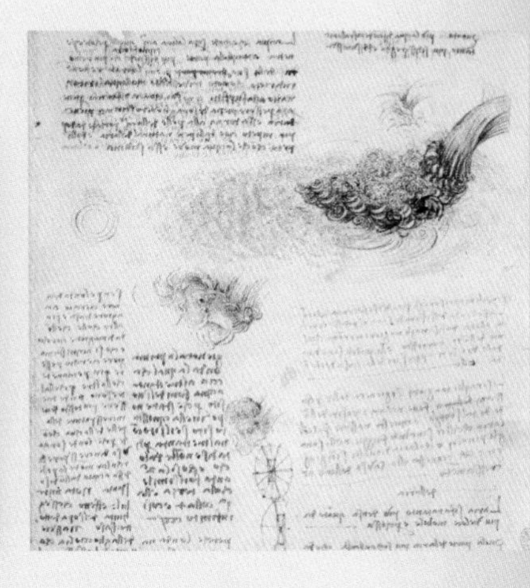

IV.
LEONARDO DA VINCI: ON THE NATURE OF THINGS
page 143

V.
LEONARDO'S GEOLOGY: A TALE OF TWO PAINTINGS
page 163

VI.
RECESSES OF THE MIND AND SOUL
page 180

BIBLIOGRAPHY.
page 213

CREDITS.
page 224

ABOUT THE AUTHOR.
page 232

FOREWARD

If Leonardo da Vinci were alive today,

he would not just be on the cutting-edge of the art scene, he would also be at the forefront of the technological revolution, the internet, as well as scientific and medical advancements. Da Vinci wore many hats, making him a true Renaissance man. Today, he would undoubtedly be a unique combination of Bill Gates, Steve Jobs, Steven Hawking, Frank Lloyd Wright and Picasso. Yet, even with this great mix, there is something else he offered—he was a great geologist. Yes, a geologist, a scientist who studies the Earth in order to decipher its mysteries.

So now that we know Leonardo's complex makeup, he's going to be our guide to Italy, revealing many of the secrets this land has held for millennia. Many ancient scientists, historians and writers have tried to understand this fascinating place, but whatever progress they made was eliminated either by time or by *scientific* positions which held that their ideas were folklore. Interestingly, it is now the advances in science and scientific testing instruments which are *proving* that which the ancients knew. In this book, Leonardo da Vinci will reveal the answers to incredible secrets, myths and mysteries about Italy that have confounded man since the dawn of time. You'll learn about:

- **radioactive waters that are good for our health, found on an island thought to have a fountain of youth**
- **the secret of living in areas with positive and negative magnetic fields, and how they influence our temperament and well being**

- a visit to the real Underworld, with scorching steam and noxious vapors
- how amber has healing properties to reduce body pain and cure throat ailments
- mysterious roads carved 30 meters into volcanic rock which crisscross necropolises
- ancient votive offerings in the form of body parts which are still found today in the form of religious candles
- sacred cave drip waters used by women to insure fertility and abundant breast milk
- the stairway to Heaven as noted in the Bible and the Led Zeppelin song
- unexplained "earthquake lights" that are thought to be UFOs
- the ominous or auspicious meaning of thunder occurring on certain days
- the oracles who forecasted the future while in a trance induced by inhaling gases emitted from the earth

Now let me tell you how I met Leonardo da Vinci. As a geologist, I only studied science. As much as I loved the subject, I still wanted to know more about art, literature, history, pretty much everything! But I didn't have the time (or the mental capacity) to understand that many subjects. Then one day I was looking at some of da Vinci's drawings and it hit me…he was a great artist, but an even greater geologist. I kept looking at his works and the more I looked, the more I understood. It was an incredible experience... as if I was discovering secrets which he had hidden in his paintings *but* were in plain sight.

While studying his paintings, drawings and notes, da Vinci became my "personal guide" to art, literature, science and engineering. He has since taken me on a magical journey—showing me the mysterious sacred rituals of the Etruscans, accompanying me to the terrifying depths of the Underworld and leaving me awestruck by the purported powers that precious gems possess.

I am now going to share these marvelous discoveries with you. As you will see in this collection of essays, the dramatic geological landscape of Italy has provided abundant Earthly inspiration for some of the greatest cultural, literary and artistic achievements of mankind.

Begin your own magical journey with Leonardo da Vinci. Open this book to any page and let him be your guide. I am sure *you* also will make some amazing discoveries!

Acknowledgements

Some of the people who have profoundly influenced me and extended their kindness, intellect and support are unfortunately, no longer on this Earthly plane. They do reside in my heart however, and are always with me. They are John E. Sanders, geologist and James Beck, art historian.

Others who have helped me in so many ways are: John Freccero, for the idea of this collection of essays, Charles Ross, Carlo Pedretti, Leo Steinberg, Charles Hope, James Ackerman, Tony Robbins, Stephen J. Gould, my super energetic Alpine guide Fulvio Casari, Burley Barthell for the title, Larissa Bonfante, Valerie Haboush, Helena Barthell, Ira Haupt II, Jay Heller and Calvin Mowatt.

In Naples, my thanks to Concetta Martina Giuliano, supervisor of the Italian translation and advisor on classical texts and languages, Bonnie Alberts, for her superb photos, Rino, Francesco Vecchione, Antonella Bianco, the team at Caffeteria Vanvitelli and last but not least, Giuseppe De Matola, whose creative talents and grand sense of humor aided me every day.

My profound gratitude to HRH Queen Elizabeth II and the staff at the Royal Library at Windsor for their generosity in making Leonardo Da Vinci's images available and preserving his legacy for future generations.

Finally, two super talents who take after Leonardo: Donald Partyka and Francesco Filippini. Donald's superb artistic-graphic talent made the text come alive in an extraordinary fashion. Francesco's artistry resulted in illustrations and a cover that were beyond my wildest dreams.

This was a massive undertaking and I pushed the limits regarding many subjects. Any errors are entirely my own, yet, if one notices an error or questions a statement and because of it, the truth can be found, it would be worth it.

INTRODUCTION

Italy:
A Land Born of the Sea

All we really know for sure about the geologic origin of Italy is that it emerged from a tropical, azure, crystalline sea. One can write about the geology of many countries with a semblance of confidence, but Italy is not just any country—inscrutability seems to be its trademark. And so, it would be foolhardy to state that anyone knows exactly how Italy was formed, not even the geologists. In fact, there are areas so complex that they can only be inferred on geologic maps. While we pretty much know the types of rocks present in Italy, how they got there is another story. There are so many conflicting scenarios and anomalies which defy geologic principles, that often, the experts can only shake their heads and go have a glass of wine after a hard day of fieldwork. With that being said, the following is a general idea of how the land was formed and a promise to let you know whenever, and if ever, it reveals its true origins.

Italy was not actually born in one piece. It is much like a patchwork quilt, with fragments of all colors, textures and ages being brought together by the potent power of the Earth. Landmasses came together from all over, sometimes pushed or pulled or just simply emerging from beneath the earth or sea. It would take 100 million years to recast, rework, recut and move all the pieces around until they could finally be stitched together. The Earth, having a discriminating aesthetic sense, produced a stunning product: a mosaic of white limestone cliffs reaching down into aquamarine waters, folded mountains of marbleized colors topped with marine fossils and volcanoes bubbling and boiling away, depositing sunny colored minerals of yellow, orange and red—a one-of-a-kind masterpiece. Over eons, these disparate bits melded together, removing

Italy's Tectonic Plates

Italy is caught in what can be described as a multi-car geologic pile-up. Giant continental plates (African, Eurasian) are pushing against each other, causing Italy to be shoved in a counter clockwise direction. Smaller plates (Adria, Anatolia) are causing earthquakes and volcanoes as they are jostled and subducted. Meanwhile, localized faults are causing the earth to move and crack open.

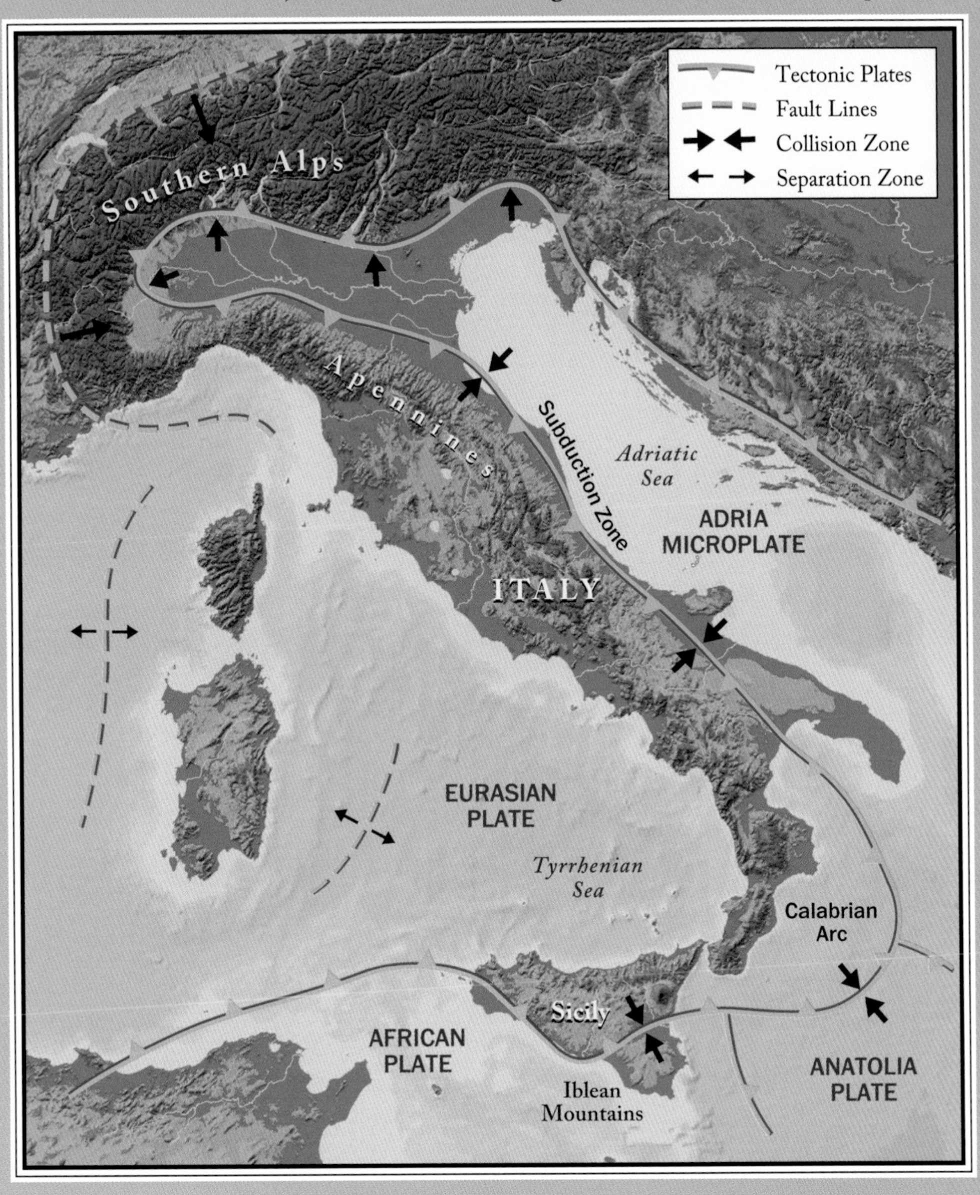

SICILIAN SUNSHINE
This finely bedded limestone was formed some 250 million years ago in the tropical Tethys Sea. It was uplifted and now forms the spectacular beach at *Scala dei Turchi* in Sicily.

many of the traces of their foreign provenance. And perhaps this is why Italy refuses to reveal its origins—it doesn't care where its parts came from. Today they are all integrated into this beautiful country and that is all that matters.

Italy was conceived during the breakup of the supercontinent Pangaea some 200 million years ago. This resulted in the global migration of all the continents. And as in any move, there was damage; some landmasses stretched the earth's crust tautly as they pulled away from each other, others, vying to occupy the same point on the globe, crashed into one another, crushing anything in their path.

An ancient sea called the Tethys formed a continuous passageway between the Pacific, Indian and Atlantic Oceans until 18 million years ago when the connection with the Indian Ocean closed, forming the Mediterranean. The closure was caused by the migration of the African plate and the collision of Arabia and Iran with Asia Minor.

EVER UPWARD

The Alps are rising every year because the Earth's continental plates are continually pushing against one another. The rocks in these mountains were once underwater, leading Leonardo da Vinci to marvel when he saw marine fossils at the top of the highest peaks.

Italy's birth began some 100 million years ago when limestone islands and rocks emerged from the ancient Tethys Sea. They would eventually form the center of the peninsula. Meanwhile, parts of Sicily were emerging from the sea and Sardinia, Corsica and Calabria were breaking away from Spain and France. If this sounds chaotic, it was, as continental plates pushed and shoved these pieces around. What would eventually happen, would be that Italy as we know it today, would be a mosaic of unrelated landmasses that were joined together by powerful Earthly forces.

Some 160 million years ago, a shallow sea, called the Paratethys, once existed in what is now the northern Alps. Then, about 145 million years ago, the Paratethys' marine sediments were thrust upward when the northward-moving African plate nudged the relatively stationary Eurasian plate. Over the next 80 million years, the African plate gathered more and more power until it pushed ahead with such force that its leading edge buckled and folded. This collision resulted in the Alps, whose turbulent birth 65 million years ago is evident by a myriad of remarkable sediments, marine fossils perched atop mountain peaks and intricately folded metamorphic rocks.

The Apennines, which form the backbone of the peninsula are some 1,200 kilometers in length. They were formed some 20 million years ago by a number of different processes which have not been discovered to this day. On the east side of Italy, these mountains are fold and thrust belt in nature and were raised by compressional forces from under the Adriatic Sea. On the west side, the mountains are fault-block, created by spreading or extension of the crust under the Tyrrhenian Sea. The Apennines are a paradox—having been formed both by extension and compression. They are called the spine of Italy because, like vertebrae, they are in constant motion. The movement is caused by the unrelenting force of the continental plates pushing against one another. Frequently, the shocks are absorbed, but sometimes, tremors are so intense that these mountains are the loci for the most violent and damaging earthquakes.

Apulia (Puglia) is the "heel" on the "boot" of Italy. It is characterized by vast areas of limestone which originated in the ancient Tethys, a shallow, tropical sea. The deposition of limestone continued for about 80 million years (all of the Cretaceous) resulting in about 3,000 meters of limey deposits, replete with coral and other marine organisms which gradually rose out of the sea in the form of islands which hugged close to the Italian peninsula. In time, because of the colliding continental plates, the sea level changed and the islands were submerged. At the end of the Pleistocene, 100,000 years ago, large expanses of limestone emerged once again, forming the base for Apulia. Subsequently, this region was covered with sand, silt and even lava, pouring forth from a deep magma chamber.

Sardinia, Corsica *and* Calabria *and* a bit of Sicily were once all part of the same land mass as Spain and France (tucked away roughly in the area from Marseille to Barcelona). Some 30 million years ago (Oligocene) the Sardinia-Corsica block (with Calabria and a bit of Sicily) broke away and rotated counter clock wise to its current location. The deep Ligurian Sea between southeastern France and the Sardinia-Corsica block could have been formed either during the rotation of the island block or by subsidence of a former land mass. About 7 million years ago the Tyrrhenian Sea began to form, first as a shallow sea underlain by extended continental crust then, after being pulled apart for 5 million years, as a deep oceanic basin.

Eventually Calabria broke away from Sardinia and made its way southeast where 13 million years ago, it stuck itself on to the Italian peninsula, becoming the "toe" on the "boot." Calabria is still heading southeastward due to plate movement as evidenced by the deep earthquakes and active volcanoes under the southeastern part of the Tyrrhenian Sea—the signatures of an active subduction zone.

Sicily is, even today, on the frontline of the collision trajectory. The situation is so complex that no one can accurately determine where the African plate intersects Sicily. We know that the African plate traverses the Strait of Gibraltar and moves east over northern

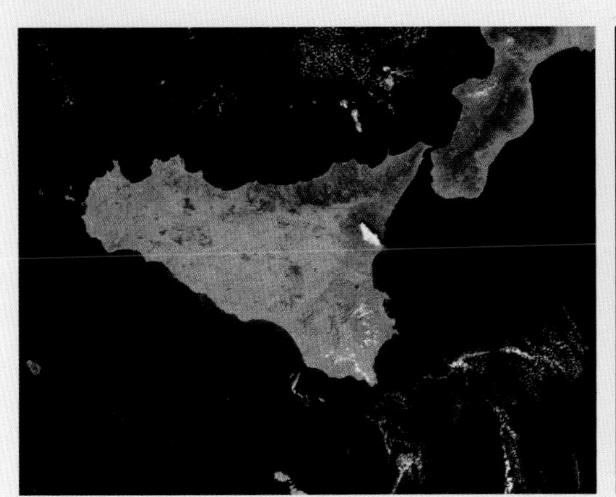

VARIED BEAUTY
The country is vastly different from north to south.
BELOW RIGHT: Venice extends over 118 small islands along the coastal plain.
BELOW LEFT: Sicily is home to a continuously erupting volcano, mountain ranges and tranquil limestone plains and beaches.

THE EARTH'S NAVEL
The ancients considered volcanoes the entrance to the Underworld, and you can see why. Italy is full of volcanoes due to its position at the intersection of several continental plates. RIGHT: Mt. Etna, a stratovolcano, has been erupting continuously for the past 3,000 years.

Africa. We also know that the islands of Lampedusa and Lampione are on the African plate, but the delineation of the plate's boundary on Sicily is still being studied. The most accepted idea is that it curves around the western edge of the island and cuts through the southeastern corner, but its precise boundary, once offshore, has yet to be determined.

Sicily itself is a patchwork of many rocks produced by the pushing, shoving and crashing of the plates. Its southeastern corner is limestone, formed in the ancient Tethys Sea. Its northeast corner was once part of the Sardinia-Corsica block which had broken off, traveled south with Calabria and stuck itself onto Sicily. The mountainous interior is an odd relationship between the Apennine chain on the mainland and the Maghrebian Mountains of Africa, leading to conjecture that Sicily was once joined to the mainland and cut off when the sea rose, or was split off from the North African range of mountains. Further, some scientists suggest that Sicily is actually being pushed toward Italy, so there is no need to

Locations mentioned in the Essays
Bellinzona
Milan
Verona
Venice
Mantua
Grotte di Toirano
Florence
Grotte di Frasassi
Elba
Orvieto
Lake Bolsena
Grotta di Pozzi della Piana
Grotta Sant'Angelo
Monte Soratte
Rome
Pirro Nord
Monte Sant'Angelo
Grotta di Scaloria
Sperlonga
Lake Averno
Cuma
Vesuvius
Grotta di Castellana
Ischia
Grotta di Pertosa
Grotte dei Cervi
Grotta Verde di Alghero
Grotta delle Veneri
Baia
Naples
Grotta di Zinulusa
Mt. Etna
N

build a bridge over the Strait of Messina because in a few million years it won't exist.

Italy is host to the only active volcanoes on mainland Europe. The volcanism is the result of subduction of one tectonic plate beneath another. Three main clusters of volcanoes exist: an arc running along the coast from the center of the country southward, northeast Sicily and around the island of Linosa. The string of volcanoes has some famous members, including Vesuvius, Stromboli and Etna.

As we look at the product of the Earth's handiwork—which is still in progress today, we can see a marvelous mosaic of color and texture. To the west of the Apennines lie the northern cliffs of the Ligurian coast with their romantic coves and deep waters. They gradually grade into the sandy beaches of Tuscany and Lazio. South of Rome, volcanoes dot the coast like stepping stones all the way down to Sicily. East of the Apennines, deltas and wetlands characterize the northern reaches, while beaches and mountains hug the central Adriatic coast. Further south, in Apulia, the waters are evocative of the Caribbean where the underlying limestone renders the sea a deep aquamarine.

The totality of the Italian landscape would in large part influence human development on the peninsula. It would impact patterns of settlement, agricultural fortunes and capacities for municipal expansion and self-defense. It is also a land which is alive with abundant natural phenomena—sacred gifts that have always enhanced the overall quality of the physical and spiritual lives of its peoples.

As this collection of essays attests, the dramatic geology of Italy has provided abundant Earthly inspiration for some of the greatest cultural, literary and artistic achievements in human history. ❂

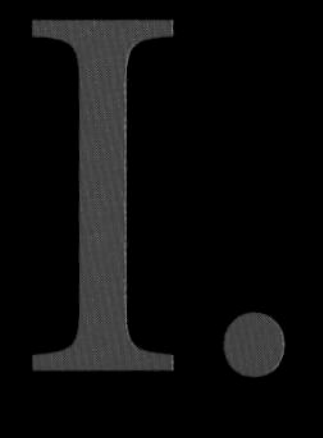

I. A Landscape Divine: Etruscans and Their Environment

The Etruscans as a nation were distinguished above all others by their devotion to religious observances, because they excelled in the knowledge and conduct of them…

—Livy, 5.1.6

ETRUSCAN OPULENCE
Thanks to the abundance of natural resources in their territory, the Etruscans lived well. The soil produced agricultural products of exceptional quality. They traded their minerals for items they desired, such as textiles, precious gems and gold, which they fashioned into exquisite jewelry.

The Etruscans made their home in a land of poisonous gases, ethereal vapors, earthquakes, volcanoes, radiation, thermal waters, magnetic anomalies and thermodynamic effects. Long considered an enigmatic population, they were actually, along with the Greeks and Romans, one of the principal peoples of the classical world. Their earliest settlements, dating to 1200 B.C., were in Etruria, an area lying between the Arno and Tiber rivers in what are now the regions of Tuscany and Lazio. Here, as mysterious and frightening as they found the land, it was also fascinating; with lava solidified into strange forms, deep gorges slicing the earth and boiling waters gushing from springs.

The same volcanic activity that made the landscape forbidding also produced valuable natural resources which allowed for the development of a society which was prosperous and aristocratic. Excellent wines and fragrant olive oils were produced from their rich soil. Mineral deposits allowed them to develop extraordinary metalworking skills and highly profitable mercantile ties. At the height of their civilization in the VII-VI century B.C., the Etruscans were the major power in Italy and competed with the Greeks for primacy in the central and western Mediterranean. Although able seafarers and adroit militarily, they never forged a unified political entity, opting instead to adopt the Greek model, a league of twelve city-states, each a loose association of towns and settlements bound by geography or common commercial interests, yet sharing a common religion and language. From the IV-I century B.C. however, this great civilization's decline was under way, as various cities were absorbed into the Roman Empire, either by having been defeated in battle or joining voluntarily for political or economic reasons.

VOLCANIC VAPORS AND SCULPTED STONES
Etruria's boiling waters and vapors emitted from the netherworld made for a mysterious landscape. RIGHT: Lava which cooled slowly formed strange hexagonal columns called "launched stones" thought to be petrified lightning bolts embedded in the earth after being thrown by a god from Lake Bolsena to punish the community which had not made proper sacrificial offerings.

Knowing the areas in which the Etruscan civilization excelled, the Romans appropriated many practices which had been discovered and refined by their defeated neighbors. One of these was their highly organized religion. From the writings of Livy (59 B.C.-17 A.D.) and other Roman authors including Seneca (4 B.C.–65 A.D.) and Pliny the Elder (23–79 A.D.), we know that Etruscan religious protocols, including the art of divination, were adopted by the Romans. In fact, Roman rulers retained Etruscan *haruspices*—priest practitioners of the *Etrusca disciplina*, the ancient art of divination—at their courts and consulted them on important military and spiritual matters.

To understand the Etruscans' outlook on religion, it is worth examining their rituals in light of the special geological, geophysical and geomagnetic properties unique to this landscape. For what sets the Etruscans apart from other early peoples of the Italian peninsula is the manner in which they partook of their environment, not only for architecture and agriculture, but for healing and ritual, viewing the totality of this dynamic and volatile landscape as a device for divination and a means to live in harmony with nature.

A Land of Violent Birth

The scenic "rolling hills" of Tuscany and Lazio belie their seismically dynamic origin. Volcanic activity in central Italy, west of the Apennines, began some 10 million years ago, resulting in the formation of the largest expanse of volcanic topography in continental Europe.

These eruptions, which continued until about 50,000 years ago, formed the volcanic hills of Tuscany, the most prominent of which is Monte Amiata. A late Quaternary (ca. 180,000 years ago) formation composed of ignimbrite (ash and pumice) strata and trachyte (crystallized material) lava domes and flows, Monte

The Volcanic Nature of Etruria

The Etruscans settled in a territory which was seismically active. Etruria is part of the largest volcanic expanse in continental Europe, (which continues down to Sicily). In the north, the Tuscan Magmatic Province consists of Monte Amiata, which today is surrounded by lakes and springs, and Lardarello, which is one of the largest generators of geothermal energy in the world. The volcanism extends south to the Roman Magmatic Province which includes Lake Bolsena, the second largest volcanic lake in the world (twice the size of the island of Manhattan) and its two islands, considered sacred by the Etruscans. The volcanoes have been eroded over time, resulting in the rolling hills of Tuscany and the sculpted landscape near Orvieto. Meanwhile, thermal waters, mud and steaming vapors are emitted continuously from the ground. The land is also dotted with buttes and hilltops which were the locations the Etruscans preferred for their settlements.

LINK TO THE UNDERWORLD
The Etruscans placed their temples near faults, fumaroles, caverns or thermal springs. They felt that these locations would provide them with a link to the Underworld. In making their offerings they hoped to appease the gods both above and below ground.

Amiata is surrounded by small lakes of volcanic origin and is still ripe with geothermal activity. Its name comes from the Latin, *ad meata*, "at the springs," for the abundant fresh water found there.

About a million years ago, a second wave of volcanism began, forming a line of volcanoes, much like a string of pearls, stretching from Lake Bolsena in the north to Vesuvius in the south. The strength of the eruptions was such that it forced the course of the Tiber River eastward to its current location.

In northern Lazio the Vulsini Volcano Complex is centered on Lake Bolsena—the second largest volcanic lake in the world after Lake Titicaca in South America. It is 81 meters deep, covers 113.5 km² and was formed some 370,000 years ago when the original volcanic caldera collapsed into a deep aquifer. The two islands in the southern part of the lake, Bisentina and Martana, were born of subsequent underwater eruptions. The most recent geological evidence suggests that volcanic activity there ceased 150,000 years ago, however historic records indicate that there may have been some sort of eruptive event at Vulsini in 104 B.C.

South of Lake Bolsena is the Cimino-Vico Complex which includes Lake Vico, a caldera lake formed some 300,000 years ago. Further south is the Sabatini Volcano Complex with the Sabatini Mountains and Lake Bracciano, a crater lake formed by the intense volcanic activity of 600,000 years ago.

In Lazio, a number of volcanic systems overlap each other, making the geology complex, but the topography very pleasing. The so-called seven hills of Rome are a result of the city being surrounded by volcanoes, the Vulsini to the north and the Alban Hills to the south, bisected by the Tiber River. The Alban Hills area has two nested calderas and numerous vents formed over a succession of eruptive cycles from ca. 600,000 years ago to ca. 45,000 years ago. Many of the vents are explosion craters, and include scenic Lake Albano, the deepest of the central Italian crater lakes and many smaller ones which dot the landscape.

The complex geologic history of central Italy resulted in a diversity of soils and rocks in the region. The volcanic material consists of debris, mud and ash, which cooled to form tuff. Deposits of the last 50,000 years or so are alluvial in nature, flowing down largely from the headwaters of the Arno at Monte Falterona and the Tiber at Monte Fumaiolo—some 30 kilometers away from each other in the northern Apennines. Erosion of the tuff has resulted in the formation of deep gorges and the exposure of an underlying layer of marine limestone which formed between 250 and 70 million years ago in the shallow waters of the Tethys Sea which once covered most of southern Europe. Evident too in these deep cuts in the earth are long periods of seismic tranquility during which dense forests thrived in the mineral-rich ejecta, only to be smothered by subsequent eruptions. These appear as black and yellow bands within the tuff. Collectively, these geological processes endowed ancient Etruria with an extraordinary landscape.

TUFF TERRITORY
ABOVE: The volcanic tuff that covered Etruria provided hilltops for secure settlements such as Civita di Bagnoregio. BELOW: It was also ideal for easy excavation of tombs or sanctuaries.

Fruits of Earth's Forge

The volcanic land provided the Etruscans with the opportunity to live well and thrive economically. The soil produced exceptional agricultural products. The tuff, deposited over a large aerial extent, provided free building material and was used for the construction of homes, temples, walls for defense and tombs. It also served as an ideal material for carving out cuniculi (channels) which collected groundwater or rainwater for domestic use. Iron, lead, zinc, copper, antimony, mercury, tin and silver were the result of volcanic activity and the subsequent metamorphism in the area. The mineral deposits of Etruria not only provided vital raw material but served as a source of wealth when traded. Tuff represents the predominant stone of the Lazio landscape from which the Etruscans excavated their necropolises. Peperino, *lapis albanus*, is a rock which originated from the Cimino Volcano near Viterbo and is also found in deposits south of Rome. It is named after the dark detrital material embedded in the stone which cause spots

SCULPTED BY NATURE
Over centuries the tuff was eroded—the end result: delicate drapes of rock (called *calanchi* in Italian) which cascade over the hillsides.

resembling pepper corns. Being durable and fire resistant, it was used for bridges and aqueducts.

In Etruria, the deep gorges resulting from the erosion of the tuff by surface water runoff made transportation and communication difficult. Yet there are many traces of ancient roads and paths, cut through the rock, which hug the sides of the cliffs. With increased economic development and wealth, trails were replaced with roadways capable of accommodating carts. Many of the cart tracks are still preserved in the roads which still crisscross Etruria. In fact, Etruscan road building is impressive from an engineering standpoint. Many roadways were dug deeply and designed for stability and efficiency. While the tuff was easy to excavate, it was not that durable. Instead of importing harder rock slabs as surfacing material, the Etruscans recut and leveled the roads as they became

ENGINEERING EXCELLENCE

Since the territory was sliced by deep gorges, the Etruscans became road and bridge builders *par excellence*. While their bridges did not survive, this one, the Roman Badia Bridge at Vulci still has the original Etruscan foundation, made of blocks of red tuff. In fact, this Roman reconstruction is similar to the original Etruscan design. This is also known as the Devil's Bridge, as only he could be capable of constructing one so high (30 m.) with such a wide span (20 m.).

worn. The deep gorges which cut through the landscape required them to be master bridge builders. While the bridges themselves have not survived, remnants of the roads leading to them and their foundations can be found in many parts of Etruria.

Fertile soils, defensible outcrops, bountiful building materials and abundant fresh water would have been enough to provide any population occupying the region with the makings for a comfortable way of life. But the landscape offered more in its ongoing geothermal activity—vapor emitting hot springs, which the Etruscans recognized early on for their healing powers.

The ancient Via Cassia passes through the "Plain of the Baths," known as such for the thermal springs that have been famous since antiquity. At Sasso Pisano, (Castelnuovo di Val di Cecina) we find the only surviving example of an Etruscan sacred thermal complex

HEALING HOT SPRINGS

The Etruscans realized the health benefits of these thermal waters and established public baths. Besides the sulphate, bicarbonate, calcium, sulphur and bromidic salts, the waters contained small amounts of radioactivity. The Montecatini Terme, founded in 1530, owed its grand success in the XX century to Marie Curie who endorsed the spa because of its radioactive waters.

LAKE BOLSENA AND ITS ISLANDS
The second largest volcanic crater lake in the world was sacred to the Etruscans. Besides being considered the umbilical to the center of the Earth, the mysterious islands that floated in its crystal clear waters, Bisentina (RIGHT) and Martana, may have been the loci for rituals and celebrations.

from the II century B.C. The intricate hydraulic system, designed to carry water from the nearby hot springs, was an extraordinary engineering accomplishment. Limestone channels conveyed the water to the pools used for bathing and to a great outdoor fountain. This thermal spa was built upon an existing structure dating from the III century B.C. Two thermal installations with pools and service facilities for bathers were found and a seal with the Etruscan inscription *spural huflunas* ("of the city") appearing on numerous roofing tiles most likely indicates that the baths were public. Abandoned for almost a century after having been damaged by a seismic event in the second half of the I century B.C., the complex, partially rebuilt, remained in use by the Romans who used these hot springs, rich in salts, for the dyeing of cloth.

Courting a Restless Earth

The Etruscans considered natural features such as mountains, lakes, springs and groves sacred. A prime example being Monte Falterona at the headwaters of the Arno where thousands of votive offerings were found, including terracotta and bronze figurines and anatomical models for healing requests. Topographic elements such as volcanoes, hot springs, steam-emitting fissures, grottoes and other clefts in the rock were of particular importance when it came to the placement of sanctuaries. Viewed as entrances to the Underworld, such openings in the earth offered a link from this world to another. Many exist even today: Lake Averno near Cuma, Lake of the Sibyl on Monte Vettore and Lake Cotilia near Rieti.

The Etruscan god Sethlans was the patron of beneficial fire, including that of volcanoes. He was subsequently known as Vulcan by the Romans and was worshipped at an annual festival on August 23 known as the Vulcanalia. The ancients called volcanoes and crater lakes *omphalos,* from the Greek, meaning umbilical. They believed that Monte Amiata, the only volcano in Tuscany, was sacred; surrounded by a forest, springs and thermal waters. The crater lake Bolsena was especially appreciated as it had two islands, Bisentina and Martana which were formed by underwater eruptions after the collapse of the surrounding caldera. While evidence of an Etruscan presence on these islands has been noted, there is no basis for the legend that they were the meeting places for the heads of the twelve tribes. In fact, archeologists have now located the meeting place, called Voltumna, not far from the lake at Orvieto *(urbs vetus).*

Overlooking the crater of Lake Bolsena between the cities of Bolsena and Montefiascone is an Etruscan hill-top settlement known as Civita di Arlena excavated by the French archeologist Raymond Bloch in the 1950s. To reach the site one has to climb through the woods, passing the natural springs of Turona. Surrounded by rocky cliffs, the area consists of remnants of dwellings and a temple.

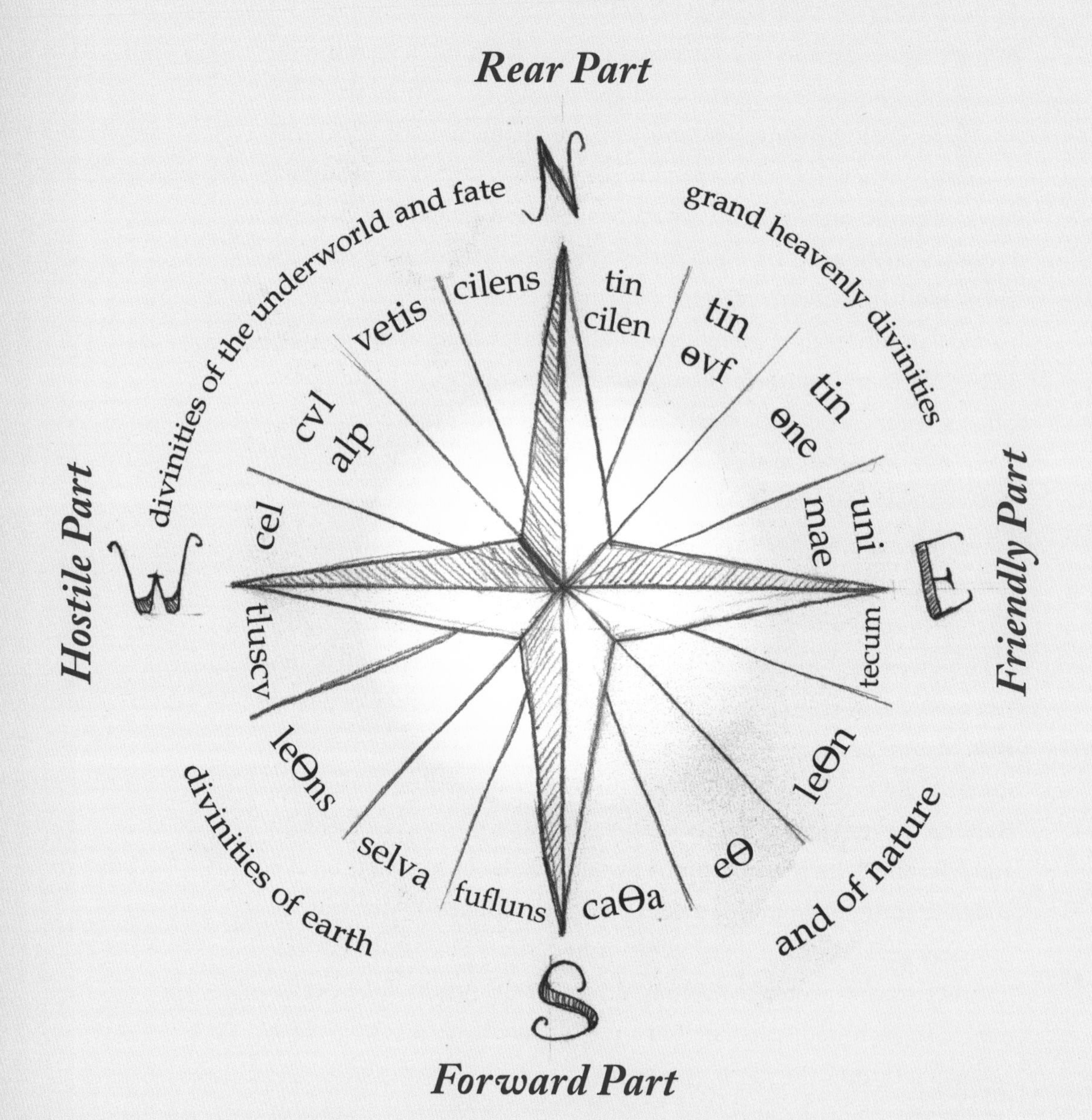

ETRUSCAN SACRED GEOGRAPHY

The cosmos was divided into 16 parts, with a god reigning over each one. The northeast quadrant was the most auspicious, while the northwest was associated with misfortune.

COROT'S ETRURIA
Jean-Baptiste Camille Corot visited Civita Castellana in 1826. He captured the charm and the warmth of the tuffaceous landscape. His choice of color and texture conveys monumentality, yet a wistful remembrance of time's passage and the people who once lived there. Today, artists still come to capture this alluring landscape.

While the temple is often referred to as the temple of Turan, (Aphrodite/Venus for whom the natural spring was named), Bloch thought it might have been dedicated to a chthonic (deity of seismic and volcanic activity) goddess because of the presence of a deep natural fissure in the rock adjacent to the temple. The narrow crevasse, of seismic-volcanic origin is about 15 meters long.

The remains of several temples dating to the VI–III century B.C. were found in Faliscan territory at Falerii (Civita Castellana). One is the Temple of Uni (Hera/Juno). The remnants were discovered in a deep valley where two volcanic gorges cross. The structure is embedded between tall, tuffaceous rocks. The temple was oriented toward the southeast and its perimeter walls, still visible, were built of tuff. Channels carved into the walls carried a local spring's water, used in ceremonies, to the sanctuary. In 1873, not far from the temple, in an area known as Fosso dei Cappuccini, two caverns were found. They contained votive offerings as well as a large basin which collected water for rituals.

It is clear from the siting of these sanctuaries and the offerings placed in them that the Etruscans understood the telluric importance of such locations. While many were pastoral, near bubbling brooks or sheltered valleys, many were truly dangerous, linked to the geologic underworld boiling just below. Perhaps it was believed that sacred sites in proximity to the netherworld might appease the gods.

CAVERNOUS ROADS

No other civilization produced cavernous roads with side walls reaching 20 to 30 meters high. The Etruscans were masters of cutting and carving volcanic tuff, but these roads, (now listed with the World Monuments Fund) with their mysterious incised symbols, are simply breathtaking. While we still do not know their function, various ideas have been put forth regarding their purpose: canals for collecting water, walkways, roads for escaping enemies. None of these ideas are feasible as some of these roads are parallel to each other, while others are labyrinthine. In many ancient civilizations a labyrinth was associated with the spiritual journey and a passage through it was the equivalent of an initiation into the mysteries of rebirth.

So the question lingers, why exert such a monumental effort to carve out a walkway or canal? The one thing these cavernous roads have in common is that they all cross necropolises, which were sacred. Given that the Etruscans were a populace that believed in ritual and held that natural areas such as grottoes, woods and springs were holy, it may make sense that these cavernous roads were used for religious ceremonies associated with the afterlife. Indeed, slicing through the rock to such a depth is like opening a pathway to another dimension.

Notions of The Sacred and Profane

The Etruscan religion was highly organized. We know from classical authors that the Etruscans possessed a vast corpus of literature, including volumes of canonical texts related to the performance of prescribed rituals and the *Etrusca disciplina*, the ancient art of divination. The latter, contained in the *Libri Tagetici*, is said to have been imparted to the Etruscans by the infant Tages, a prophet born of the earth. Emerging from a freshly plowed furrow in a field at Tarquinia, Tages revealed his wisdom to Tarchon, founder of the city, requesting that he write it down and share it with his people. Still other religious tracts were attributed to Nymph Begoe (*Vegoia* in Latin), a prophetess who is thought to be the source of *Libri Vegoici*, the books on lightning that were kept in the Temple of Apollo at Rome. Unfortunately, little original Etruscan literature has survived. What have come down to us are numerous inscriptions, the earliest of which are dated to the VII century B.C. Among the thousands of funerary epitaphs, boundary markers and carved objects are several key texts which have shed light on aspects of Etruscan religion.

From these documents and the images painted in tombs and engraved on mirrors, as well as commentaries by later authors, scholars have been able to piece together much of the Etruscan pantheon which was presided over by

ETRUSCAN PRIEST
The *haruspex/cepen* was responsible for the welfare of the community. He not only presided over religious ceremonies, but interpreted signs from the gods and provided secular and military council.

the supreme deity, Tin. He is the Etruscan equivalent of the Greek Zeus and the Roman Jupiter and possesses a strong affinity to Tarhunt, the Hittite storm god—the bearing of lightning bolts one of his key attributes. Important too were Tin's consort, Uni (Greek Hera/Latin Juno); Turan (Aphrodite/Venus) goddess of love; and Sethlans (Hephaistos/Vulcan), the god of beneficial fire, the forge and volcanic activity.

The Etruscans developed a well-defined system of sacred geography which was integrated into their daily lives. All ritual and religious observance was based on the division of celestial and terrestrial space and sacred and secular affairs had to be coordinated with it. They believed that Heaven and Earth were crossed by a north-south axis, *cardo*, and an east-west axis, *decumanus*. Each quadrant was then quartered, resulting in 16 subdivisions, each ruled by a specific deity, and each with varying levels of auspiciousness. The word auspicious comes from the Latin *haruspex/auspex*, one charged with divination.

The priests had to decipher signs coming from the gods depending on the position from which they emanated in the sky. The east was a favorable location because there, the deities most sympathetic to man had chosen to dwell. The northeast was the most auspicious and promised good fortune. In the south the gods of Earth and nature ruled. In the west, the terrible and merciless gods of the Underworld and of fate dwelled. The quarter between north and west was considered the most inauspicious.

The Etruscans even evolved a system of town planning based on these religious concepts, which were likewise reflected in the elaborate ritual prescribed for the foundation of a new city. In Etruria a town laid out in accordance with the sacred rules was considered a minute portion of the cosmos, harmoniously integrated with an all-embracing order governed by the gods.

Before ground would be broken for a new town or settlement, an Etruscan priest, or *cepen*, wearing his conical hat (which survives today in the form of the Bishop's mitre) and holding his lituus (the Bishop's crosier), would fix the north-south and east-west lines by the sky, then solemnly mark out the cross of the *cardo* and the *decumanus* on the ground. Only then could the first streets be laid and the first buildings erected. Civic boundaries were often marked by a ritually plowed furrow or a cippus, a boundary stone. Such boundaries were presided over by protective deities. Thus, this notion of well-defined ritual space became integrated into the daily lives of the populace.

Messages from Heaven

The chief priests were important to the welfare of the community and always sought to ensure a peaceful interconnection between Heaven and Earth. Living in such a geologically active place, the priests had to evaluate natural occurrences—both benefic events as well as catastrophes in order to interpret the signs being sent from the gods. The Etruscan priests were persons of extreme intelligence who had undertaken a long and laborious course of study which began at an early age. They were tutored by the elder priests and not only studied sacred teachings, but were required to master astronomy, meteorology, zoology, ornithology, botany, geology and hydraulics.

There were three primary means of divination used by the priests—the interpretation of lightning and thunder: ceraunoscopy/brontomancy, the scrutiny of entrails (the liver in particular) of sacrificed animals: hepatoscopy and through observations made on the flights, cries and behaviors of certain birds, namely ravens, crows, falcons, owls and eagles: ornithoscopy/augury. Signs from the gods could manifest in two forms: voluntarily (*ablativa*) such as lightning and thunder, and those asked for (*imperativa*), such as the reading of entrails and bird flight.

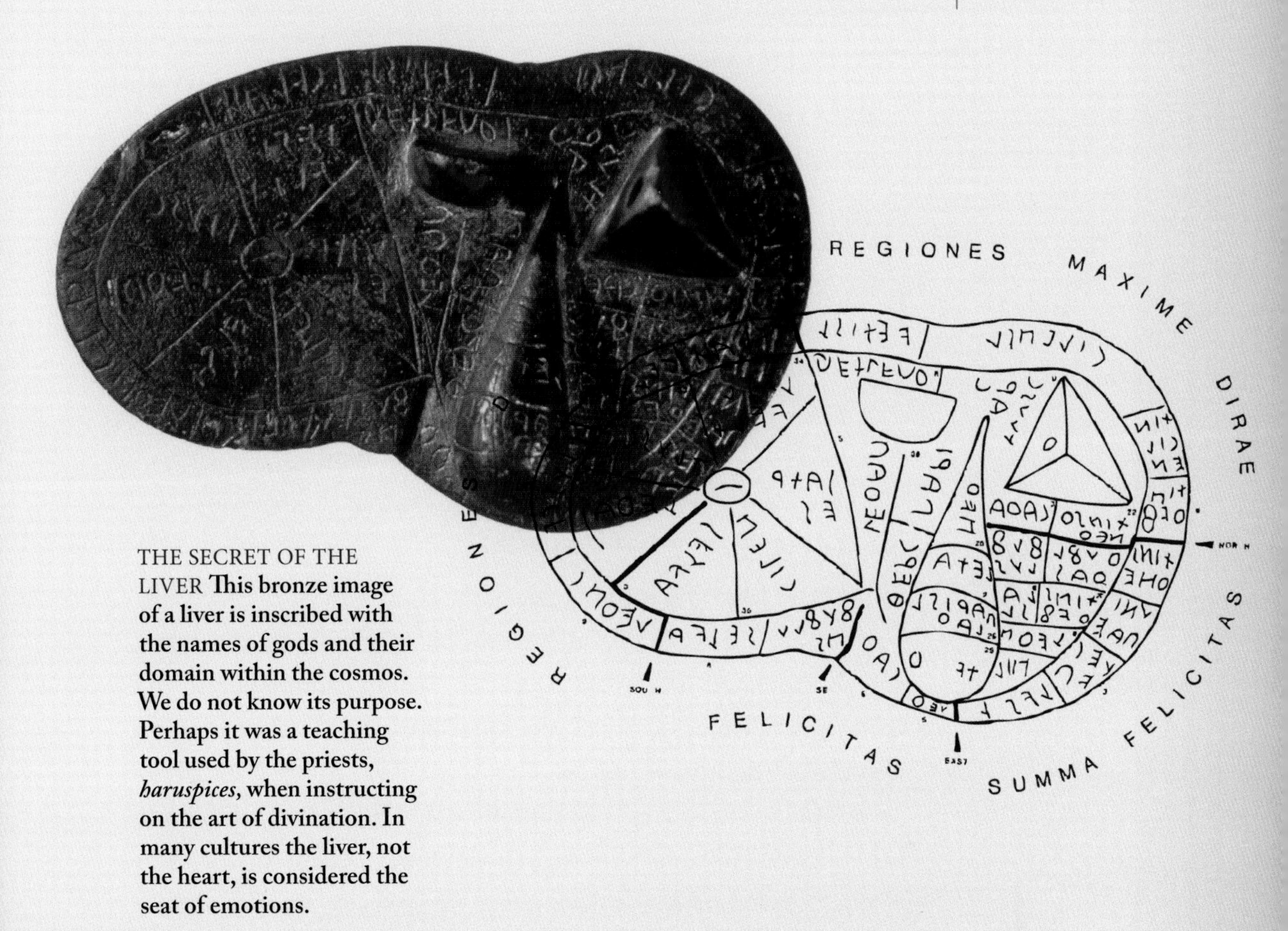

THE SECRET OF THE LIVER **This bronze image of a liver is inscribed with the names of gods and their domain within the cosmos. We do not know its purpose. Perhaps it was a teaching tool used by the priests, *haruspices,* when instructing on the art of divination. In many cultures the liver, not the heart, is considered the seat of emotions.**

The Etruscans consulted a brontoscopic calendar (*bronte* meaning thunder in Greek), a thunder-based daily divinatory calendar covering the 12 lunar months (this was a common method of time keeping in antiquity) of the Etruscan year which started in June. Although the original Etruscan text has not survived, the calendar has been preserved in a VI century A.D. Greek transcription. It was like a modern *Farmer's Almanac,* containing a wealth of social, agricultural, religious and medical information and was specifically designed to be used as a reference for priests interpreting thunder.

Here are some examples of the daily forecasts:

August 19: If it thunders, the women and the servile class will dare to commit murders.

October 21: If it thunders, there will be coughing sickness and heaviness of the heart.

November 17: If it thunders, there shall be plentiful fodder for the flocks.

February 4: If it thunders, men shall be troubled not only in their countenance but also in their very minds.

March 11: If it thunders, it signifies torrential rain and the creation of locusts.

According to Jean MacIntosh Turfa of the University of Pennsylvania Museum of Archaeology and Anthropology, who has undertaken extensive research on the calendar, the Etruscans understood the connection between seasonal precipitation and animal health and husbandry. She notes that the Etruscans were aware that excessive rainfall in the spring and summer could dramatically increase outbreaks of Fasciola hepatica (liver fluke, a parasite) in their livestock. As evidence for the disease was visually manifest in the liver of an infected animal, the brontoscopic calendar would have had a pragmatic use in forecasting potential risk to animal and human welfare.

Reading the entrails of sacrificed animals was a divinatory practice that was highly refined by Etruscan priests *haruspices.* One of the teaching tools they might have used was a life-sized inscribed bronze sheep's liver from Piacenza, Italy ca. 100 B.C. The liver is divided into sections and inscribed with the names of 51 gods (some mentioned several times). The outside perimeter is divided into sections corresponding to the 16 regions of the heavens as defined in the Etruscan religion. While the exact methodology or use is not known, scholars think that based on the location of a mark on the liver, a priest could foretell the future or even bend it to his will.

BIRD AUGURY
The birds depicted in tombs are symbolic of the journey to the afterlife. We do not know much about the intricacies of bird augury as practiced by the Etruscans, but we now know that birds have an internal compass of magnetite that allows them to determine direction.

With regard to divination by the behavior of birds, Pliny and others make numerous references to such observations, but it remains unclear just how the Etruscans may have interpreted these omens. While both falconry and augury are depicted in a number of Etruscan tomb paintings, these images represent the passage to the afterlife and do not give us insight into the practice of augury.

Among Pliny's writings (*Naturalis historia*, x), he mentions several types of bird omens. We are told that *"seeing the crow near the summer solstice is most inauspicious and that ravens are the only birds that seem to have any comprehension of the meaning of their auspices."* With regard to the horned owl, Pliny states that *"it is especially funereal, and is greatly abhorred in all auspices of a public nature. Hence it is looked upon as a direful omen to see it in a city, or even so much as in the daytime. I know, however, for a fact, that it is not portentous of evil when it settles on the top of a private house."*

As Erika Simon, professor emerita of Würzburg University, has pointed out, all three means of divination have their roots in Anatolia and the ancient Near East. Inscribed clay models of livers for divination similar to the Piacenza Liver are known from the mid-II millennium B.C. at the Hittite site of Hattusa in Turkey and from the Babylonian site of Sippar in southern Iraq. Bird augury was practiced by the Hittites as early as the XIII century B.C. as evidenced by a number of bird oracle tablets which were found at Sarissa (modern Kuşaklı) in northeastern Turkey.

Indeed, Herodotus tells us that the Etruscans traced their origins to a migration of peoples from the Lydian region of Anatolia (modern-day Turkey) in the wake of a great famine. While scholars have long sought to determine the origin of the Etruscans—whether they are a population indigenous to Italy, of foreign extraction, or a blend of both—archeologists and genetic researchers now agree that the Etruscans developed in-situ. As research continues, scientists are trying to explain the Anatolian genetic material which shows up in the modern population of Tuscany (ancient Etruria). Marco Pellecchia of the Università Cattolica del Sacro Cuore, started with a genetic study of bovine populations (*Bos taurus*) in Italy. Six breeds unique to the Etruscan region exhibited mitochondrial DNA (mtDNA) variations common in breeds in Anatolia and the Near East but which are lacking in the rest of Italy or in Europe. Tests also showed that this genetic signature is paralleled in modern human populations from Tuscany, which bear genetic affinities to those from Anatolia and the Middle East. Subsequent studies however, attribute this genetic mix to the influx of immigrants, who also brought livestock, into Etruscan territory.

Signs of The Divine in Nature

Beyond the gases, steaming vapors and hot springs the Etruscans saw in this seismically charged environment, it is home to some geological and meteorological phenomena so rare and unexplainable, they would have had to be attributed to the divine.

According to Seneca (*Naturales Quaestiones*) the ancients knew of a type of lightning called *fulmen hospitale*, which they believed man could invite Jupiter to send or "draw down." Once seen, it would then be interpreted by a priest who specialized in lightning, the *haruspex fulgurator*. Perhaps Seneca was referring to cloud to ground lightning which is one of the three general types recognized today, the other two being cloud to cloud and within a cloud. Pliny elaborates even further: "*The Tuscan books inform us that there are nine Gods who discharge thunderstorms, that there are eleven different kinds of them, and that three of them are darted out by Jupiter.... The Etrurians also suppose, that those which are named Infernal burst out of*

LAND OF LIGHTNING
The Etruscans observed different types of lightning and categorized them according to color, form and the location from which they originated and returned. This information was interpreted by the priests as many natural occurrences were considered messages from the gods.

the ground; they are produced in the winter and are particularly fierce and direful, as all things are which proceed from the Earth, and are not generated by or proceeding from the stars, but from a cause which is near at hand, and of a more disorderly nature."

Pliny's observation suggests that the Etruscans may have witnessed ground to cloud lightning or possibly lightning generated by volcanic activity or the recently accepted phenomena called "earthquake lights." Ground to cloud lightning favors terrain characterized by isolated small peaks at high elevations. It also occurs when intense electric charges concentrate on the tip of a structure during a storm. Since many of their cities were constructed on hilltops it is likely the Etruscans witnessed numerous strikes as their buildings would have been easy targets. Interestingly, legend has it that Tarchon planted white bryony (*Bryonia alba*) vines around his abode to protect against lightning bolts.

Volcanic terrain is also conducive to lightning production. Gases, ash and pulverized material in the air create a plume of highly charged particles which provide the perfect electrostatic conditions for lightning. In fact, recent studies by Steve McNutt of the Alaska Volcano Observatory show that a new type of volcanic lightning might have been discovered with some bolts reaching as high as 3 km. While Martin Uman of the University of Florida is not convinced, he readily adds, *"any kind of volcano lightning is super-gorgeous, it's one of our best natural phenomena."*

The Etruscans, in all likelihood, categorized as lightning what are today called "earthquake lights," strange illuminations of various forms and colors, rising from the ground, often with a crackling sound. The earliest known report of these lights dates to 373 B.C. when the Greek cities Helice and Buris were destroyed by an earthquake accompanied by *"immense columns of fire"* according to Seneca. In another reference, Pliny the Elder (*Natural History* II. 138, 52) describes the god Satre who occupies the malefic northwest

quadrant of the Etruscan cosmos as follows: *"frightening and dangerous god who hurls his lightning from his abode deep in the Earth."* These phenomena were long held to be folklore, but we have finally acknowledged their existence today, although reliable pictures are scarce and data come mostly from eye witness accounts. Somehow, subterranean stress is causing a luminous effect in the atmosphere before or after an earthquake and it can be visible even hundreds of kilometers from the epicenter. This phenomenon still remains scientifically unexplained.

Over the past several decades, a NASA physicist, Friedemann Freund, has been working assiduously to account for these lights. His work initially met with much skepticism but he is gradually gaining support for his fascinating hypothesis. He states that igneous and metamorphic rocks contain electric charge carriers, which have been overlooked in the past. These rocks are full of flaws, one of them being the millions of oxygen atoms which are short one electron, bound by peroxy (O_2) bonds. When the bond is broken, the result is a pair of "holes" of positive charge or p-holes. The "awakening" of dormant p-hole charge carriers by the stress of an earthquake turns the rocks momentarily into p-type semiconductors (where electrical conduction is due chiefly to the movement of positive holes). Once the p-holes are generated, a steady flow of current through the rocks causes them to sparkle and glow, produce electromagnetic and positive ion emissions and mid-infrared radiation. They may also account for radio noises, disturbances in the upper atmosphere and animal and human behavioral changes.

The phenomenon of ball lightning, free-floating volumes of ionized air that detach themselves from the ground is known today as "an unsolved problem in atmospheric physics." Eyewitnesses report ball lightning entering into rooms through windows without leaving a trace or any cracks in the glass, or have entered through telephone jacks and electric sockets. While drifting through the

THE ART OF INTERPRETATION
Etruscan priests had to guide and protect the community by interpreting the signs being sent by the gods through various forms of lightning.

air, the orbs of light reportedly produce a faint hissing sound and explode with a bang after a few seconds, leaving behind a smell of ozone. They appear before or during large thunderstorms and seismic activity. During earthquakes, (according to the p-hole hypothesis) these plasma balls may detach themselves from the ground when p-hole charge clouds arrive at the Earth's surface creating highly charged electric fields at the ground-to-air interface. The charge will be particularly high on hills, ridges and mountain peaks.

The Etruscans cited many colors of lightning, *manubiae albae* (white lightning), *nigrae* (black), or *rubrae* (red). They recognized those which came from Tinia (3 kinds) from Uni, Minerva, Sethlans/Vulcan, or Laran/Mars. Finally, Satre, especially in winter, could generate chtonian thunder from the depths of the Earth. We now know that six types of luminous phenomena have been documented: seismic lightning (much like sheet lightning), luminous horizontal or vertical bands in the atmosphere, luminous orbs or globes floating in the atmosphere, fire tongues creeping along the ground like *ignis fatuus* (will-o'-the-wisp), seismic flames emerging from the ground and coronal discharge-like lights.

Since the Etruscan territory was characterized by a geologic and meteorological makeup which was conducive to the various types of luminous phenomena which we are just accepting today, it is more than likely that the Etruscans saw some wild and amazing light shows. What is more impressive is that they observed, categorized and sanctified these phenomena in a way that served them and gave homage to the forces of nature.

Given the significance Etruscans placed on lightning, they may have been aware of some of its earthly by-products, namely fulgurites and lodestones. Often referred to as "petrified lightning" a fulgurite (*fulgur* is Latin for lightning) is a tube of vitrified sediment formed when a lightning bolt strikes the ground. As the bolt continues its path into the earth, the extreme heat from the charge vitrifies the sediment through which it travels, creating a root-like web of misshapen tubes and lumps of fused sand. The deepest evidence of this vitrification process has been found 15 meters below ground. While fulgurites can be up to several centimeters in diameter and several meters long, most are quite small, such that they can fit in the palm of one's hand. They can also vary in color from black to green, brown, white and transparent depending on the composition of the sand.

A fine example of a fulgurite was found among the offerings at a Zeus sanctuary during recent excavations atop Mt. Lykaion in the Peloponnese. According to geologist George H. Davis of the University of Arizona, the fulgurite, which is some 2.5 inches (6 cm.) long, was found near a bronze hand grasping a lightning bolt, likely part of a Zeus figurine.

MAY THE GODS PROTECT US
Aplu, god of light and weather is depicted on this coin (III century B.C.) from Popolonia wearing a laurel headpiece. The mark of value on the reverse side is blank.

PRODUCED BY LIGHTNING Magnetite (HERE AND BELOW RIGHT) was used in compasses and affected the geomagnetic fields of the earth in Etruria. Fulgurites (BELOW LEFT) were formed when lightning hit the ground and vitrified the sand grains fusing them together into irregular shapes.

While the Greeks may have placed fulgurites among their votive offerings, the Etruscans did not. For they believed that anything struck by lightning must be blessed, buried and never again touched. When lightning struck the ground, the *haruspex fulgurator* was summoned to inspect the site. He then carefully collected the soil which had been struck, buried it and enclosed the area with a ring of stones called a *puteal* (Lucan, *Pharsalia* I, 606). The *haruspex fulgurator* would make an offering of a two year old sheep (*bidens*) and the consecrated area, called a *bidental*, was marked with a monument which could not be desecrated. If a human was struck, his body was buried in-situ, with the presiding priest following a specific protocol, including prayers, offerings and placing a monument at the site. The practice was adopted by the Romans. Ancient lightning markers can be seen in the Capitoline Museum and a *bidental* with a *puteal* oriented northeast-southwest, in reference to the chthonic gods governing the subterranean world, was recently discovered in Todi, Italy.

Magnetite was also an important resource for the Etruscans. In fact, Monte Calamita (Magnet Mountain) on the island of Elba was largely composed of the mineral. Legend has it that the compasses of passing ships would point to the mountain rather than north. Lodestones, bits of magnetite that have been naturally magnetized, possibly by lightning strikes, were first noted in the VII B.C. by the Greek philosopher Thales of Miletus. The term magnet is derived from the Anatolian region of Magnesia, where the mineral is found in abundance.

Ancient peoples found that whichever way magnetic materials were turned, they would always revert to the direction of the Earth's geomagnetic field. This field is thought to be largely generated by the movement of conducting material inside the Earth's core, with the rest produced by electric current flowing in the ionized upper atmosphere and telluric (within the Earth) currents flowing

Geomagnetic Map of Italy and Etruria

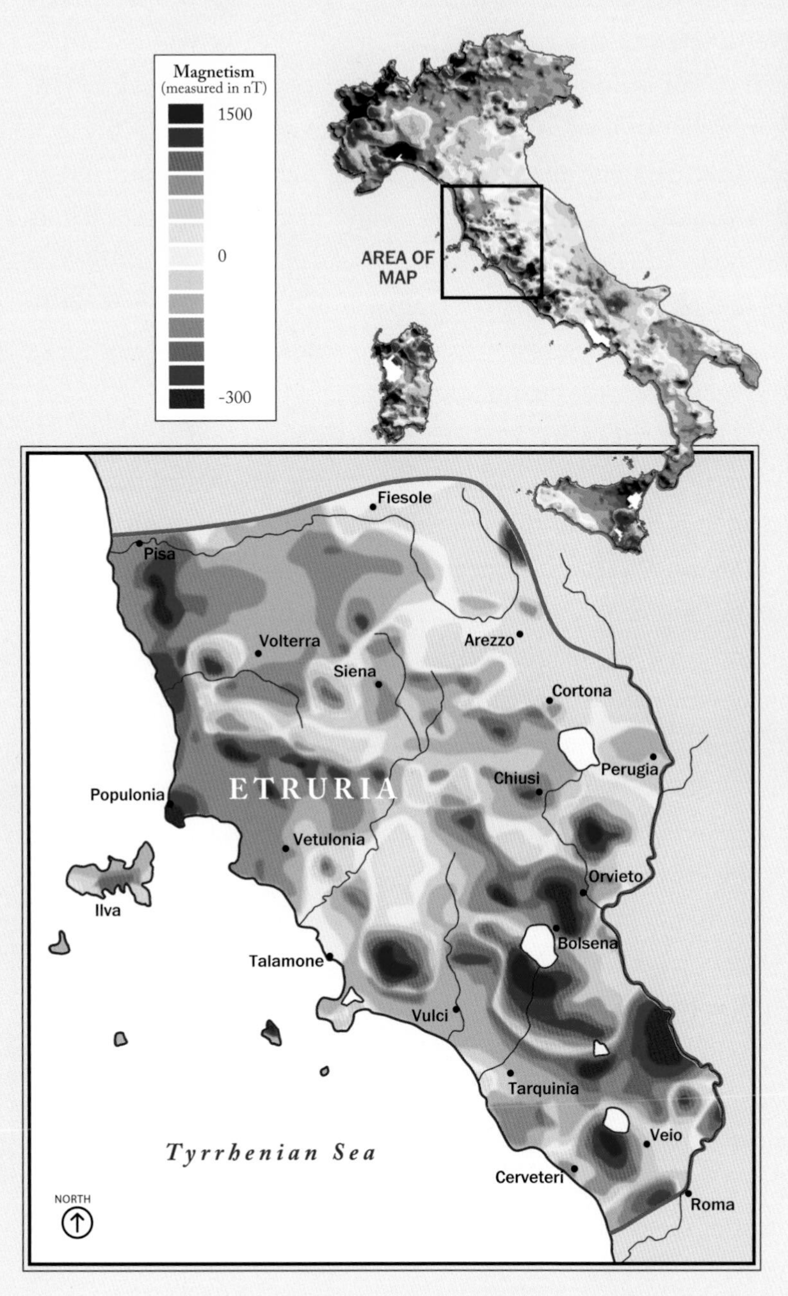

This geomagnetic map of Italy shows the diversity of the magnetic profile of the country: the blue sections being geomagnetically negative and the copper colored areas highly positive. The magnetic field strength is shown in nanotesla (nT). There are many reasons why an area would be more positive; the most common being the nature of the rocks themselves, with those more highly mafic (containing magnesium and iron) and those intrinsically magnetic generating positive geomagnetism. As can be noted on the detailed section, the Etruscans settled in an area of negative geomagnetism (blue areas). Interestingly, in regions surrounding volcanoes, which have highly positive geomagnetic fields (due to the rock types), the Etruscans placed their cities away from these areas (as seen in Bolsena and Veio). In the case of Cortona, they placed the city in an area of neutral magnetism, on a high promontory needed for defense, yet avoided an adjacent isolated area of positive magnetism. At Cerveteri, which was a vital port city, they located the settlement in an area as close as possible to neutral geomagnetism.

in the crust. There are also local anomalies produced by mountain ranges, ore deposits, faults, and in our day and age, trains, aircraft, power lines, etc. The Etruscan land was full of geologic features which could produce magnetic anomalies, areas of magnetism either higher or lower than the average magnetic field of the area. A positive magnetic anomaly is a reading that exceeds the average and is usually related to more strongly magnetic rocks, such as mafic rocks (the volcanic rocks in Etruria would fall into this category, i.e. basalt, scoria, rich in magnesium and iron) or magnetite bearing rocks. A negative magnetic anomaly is reflective of rocks which have low or no magnetite or iron.

Scientists have been laboring for decades to find a correlation between magnetic fields and health and have discovered that external magnetic fields can affect the biological systems of both animals and humans. These fields can exert a positive or negative influence on organisms depending on their intensity, frequency, orientation, exposure time and origin. With this in mind, we can look at the geomagnetic nature of Etruria on a map of Italy's natural magnetic field. It provides an unprecedented view of the magnetic signature of the geologic features in their regional setting. The map shows the areas where the natural magnetic field is positive or negative, with the Apennine chain marking a distinct boundary between two magnetic domains. According to Massimo Chiappini of the Istituto Nazionale di Geofisica, the magnetic field is primarily negative on the Tyrrhenian side, continuing north to the Po River plain, while on the Adriatic side, it is generally positive.

The map codifying magnetic variation on the Italian peninsula can now be combined with medical research data. Negative magnetic fields are now known to be calming and contribute to an overall sense of wellbeing, while positive magnetic fields have a stressful effect and can increase pain due to their interference with normal metabolic function.

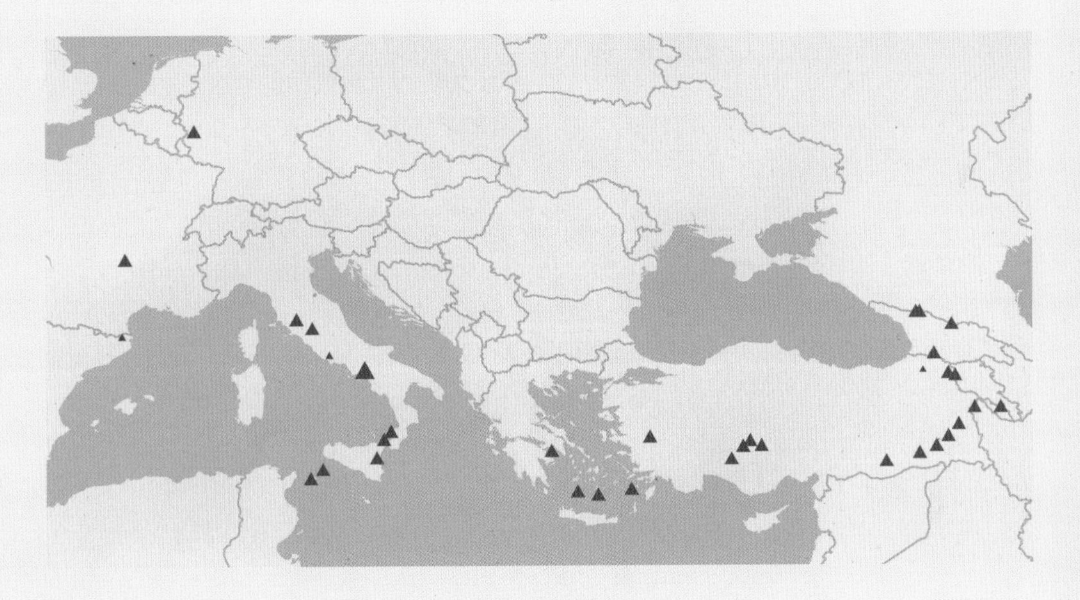

FOLLOW THE VOLCANO
A string of volcanoes runs from Turkey through Greece and into Etruria. While the Etruscans developed in-situ in Italy, people from throughout the Mediterranean followed "the volcano route" as they immigrated into Etruria.

It is interesting to note that the Etruscans established their cities in areas of highest magnetic negativity: Orvieto (Velzna), Chiusi (Clevsin) Perugia (Perusna) Tarquinia (Tarchna), Volterra (Velathri) and Vulci (Velch).

In short, the Etruscans and those venturing into Etruscan territories perhaps to luxuriate in the region's famed thermal baths might have felt better, plain and simple. Such noticeable improvements in health—while now known to have Earthly causes—may just as well have been attributed to the favorable will of the gods and the powers of local *haruspices* to conjure it for the benefit of their patrons and the population at large.

In Search of a Salutary Life

With hot springs, fumaroles, gases, minerals, magnetic anomalies, fractures, faults and crater lakes, the Etruscans must have felt that the Earth was alive. And, given their penchant for observing, organizing and interpreting natural phenomena, we must ask the question "did the geology of their land have an impact on their religion?" Most likely it did. For the Etruscans' achievements and religion cannot be divorced from their land. Their mineral and agricultural resources allowed them to become wealthy and powerful. Their water, both natural and thermal, afforded them a salutary lifestyle. It is easy to understand how the Etruscans may have attributed a divine origin to the region's natural endowment, which would not have been understood in scientific terms at the time their religious practices were adopted but are being slowly confirmed by the scientific techniques available today. In fact, their ideas still resonated 2,000 years later when the Neapolitan philosopher, Giambattista Vico (1688-1744), studied the divinatory practices of the Etruscans—in particular, that which pertained to thunder. He proposed the revolutionary idea that the sound or "voice" of thunder was the voice of God. His Etruscan inspired premise influenced generations of writers, philosophers and archeologists. ❂

CUNICULI

The Etruscans displayed a profound knowledge of geology and hydraulics. They trained expert hydraulic engineers who knew how to find subterranean water, bore wells, dig water channels, supply drinking water to the towns and install irrigation and drainage systems in the fields. They also developed specialized techniques for constructing subterranean corridors and tunneling into mountains. As a result, many types of excavations have been found, serving a myriad of purposes among which are: aqueducts, drainage, water collection, diversion works, sewers, outlets of lakes, passages, places of worship, animal shelters, quarries and tombs.

In a large part of Lazio (especially in the south) volcanoes were responsible for the local variant of tuff called cappellaccio. Often situated just below the humus, the cappellaccio stratum is impermeable to water. To drain stagnant waters, obtain cultivable land and assure its irrigation, ancient inhabitants often undertook large-scale projects, leading to a proliferation of channels, called cuniculi, which long puzzled specialists. They have now been dated to the V–IV century B.C. and are attributed to the Etruscans. Prof. Franco Ravelli has researched the origin of these structures, mapped them throughout Lazio and has also located a network of them surrounding the Capitoline Hill which are visible today.

The cuniculi are rectangular tunnels two meters high and a half meter wide. Communication between the cuniculus and the ground surface is provided by a series of narrow shafts spaced at a distance of 40 meters from one another. The cuniculi, while probably not longer than 300 meters each, spread out in Lazio for hundreds and possibly thousands of kilometers. They were excavated almost exclusively for the purpose of obtaining pure water that had been made suitable for drinking by the filtering action of the earth on rainwater. The tuff was ideally suited to this design as it was easy to excavate, did not collapse and was impermeable. Interestingly, in subsequent centuries, when a troubled history prevented the inhabitants from maintaining the channels, the spread of marshlands and malaria transformed these regions into unhealthy isolated areas, feared by all travelers until the beginning of the XX century.

CUNICULI IN ROME
The Etruscans completed many civil engineering projects in Rome, including the Cloaca Maxima, Rome's first sewer system. They also built cuniculi which collected rain and groundwater and channeled it for potable use as well as drainage and runoff control.

HEAVENLY HILLTOP HABITATS

Today, traces of dozens settlements that once dotted the landscape can still be found. It is clear that the Etruscans had a preference for hilltops as evidenced by cities such as Pitigliano, Sorano and Sovana. They are surrounded by steep volcanic canyons and many ancient sacred pathways. To access to these defensible outcrops of layered volcanic tuff, the Etruscans carved circuitous roads which snaked up the cliffside. Remnants of roads between settlements, which were punctuated with important shrines and boundary markers, can still be seen crisscrossing the region.

One of the most geologically fascinating cities is Orvieto, which was settled as Velzna by the Etruscans in the IX-VIII century B.C. It is situated on the flat summit of a large butte of volcanic tuff. The site of the city is among the most dramatic in Europe, rising above the almost-vertical cliffs that are rimmed, at the top, by defensive walls built of the same stone. Settlers have always dug beneath the city for diverse reasons; to create temples to obscure divinities, to find water, to extract building material for houses and for tombs. The result of over 3,000 years of excavating is a honeycomb of about 1,200 caves, passages, quarries and wells. Today, the area is mainly privately owned and used mainly as wine cellars as the constant temperature of 55° F. preserves wine perfectly.

MAGNIFICENT ORVIETO

Located atop a butte of volcanic tuff, Orvieto majestically dominates the landscape. At the foot of the butte is the Etruscan necropolis called "Crucifix of Tuff" which dates from the VI century B.C. and contains a hundred or so chamber tombs laid along a rectangular street grid. BELOW: Joseph Turner, in his 1828 masterpiece, captures the ethereal beauty of this ancient Etruscan settlement.

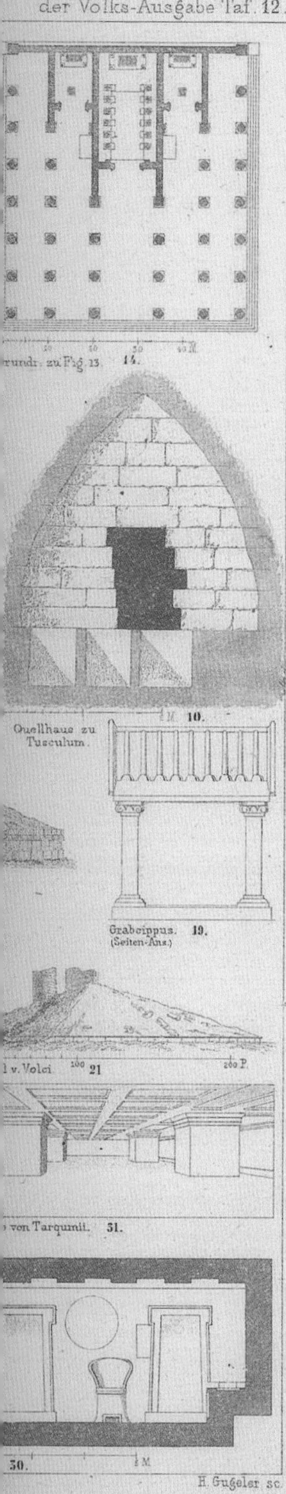

TOMBS

Geology was an important factor in tomb design as it dictated the form, location and decoration which could be used. Tombs could be built out of blocks of tuff or carved out of the ground, as the resulting subterranean rooms were resistant to collapse. The tuff also provided a surface which could be frescoed or incised with architectural detail. This was not possible with a tomb that was excavated out alluvial material as the walls were irregular and the rock subject to more rapid deterioration. The Etruscans made it a point to actively search out locations with underlying tuff, as this was the preferred building material for their tombs.

Burial in a simple trench was practiced during the Iron Age and continued throughout history. However, the suitability of tuff revolutionized the options available for burial and tombs reached monumental proportions. The Regolini-Galassi tomb (600-650 B.C.) found at Cerveteri is representative of this shift in burial practice and tomb design. The tomb contains a corridor cut out of tuff which is 120 feet long and 6 feet wide leading to the burial chambers. The upper portion is built with square tuffaceous blocks. It is covered with a tumulus (burial mound) resting on a circular wall of tuff blocks which is 150 feet in diameter.

Burial mounds often contained several tombs. They were designed to resemble homes, with a series of rooms which would hold the sarcophagi of family members and those goods which would be needed in the next world. In fact, the items found in the burial chambers also reflected the mineral wealth of Etruria; silver, copper, bronze and imported gold were fashioned into extraordinary objects of inestimable value.

EVERYTHING ONE NEEDS IN THE AFTERLIFE
Tomb of the Reliefs in Cerveteri (late IV century B.C.) is carved out of tuff and painted with stucco reliefs of household items, pets and anything needed in the next world. Even round stones, resembling marbles, to be used for recreation in the next life, were found among the grave goods. Along the walls are the actual tombs, which are carved to look like beds with pillows. The idea was to have the tomb resemble a room in one's home. The tuff allowed for easy sculpting of the architectural details such as the sloping roof and supporting pilasters.

A vast expanse of tuff was ideal for the construction of a large necropolis, called the "City of Tuff," which can be seen today between Pitigliano, Sorano and Sovana in the southeastern part of the Province of Grosseto. What is extraordinary about this location is that the tombs are connected via a series of walkways known as the "Cavernous Roads," excavated in some areas to a depth of 30 meters below the ground surface. The tombs in this area, Sirena, Tifone and Ildebranda exhibit some of the finest forms of Etruscan funerary architecture.

SACRED STONES
Sepulchral altars with a "magical" stone were perhaps the imageless symbols of a god or of ancestors. Spherical stones stood at the entrance to the tomb or on top of the burial mound. Were they symbols of the dead, a place of sacrifice, or both?

Stone was also used for sarcophagi with peperino,tuff and limestone being used as well as terracotta. Another material, "fetid stone," a limestone imbued with a strong odor of onions caused by the sulfur in the underlying ground water, was a favorite of the Etruscans for funerary sculpture and sarcophagi. Fortunately once exposed to the air, the unsavory fumes of the sulfur eventually dissipated.

TOMB OF THE LEOPARDS

In Tarquinia, located in the Monterozzi Necropolis 480-450 B.C., this fresco shows male and female banqueters being served by two nude boys carrying serving implements. The leopards above are surrounded by a marvelous checkered pattern ceiling. The purpose was to have the tomb resemble a home. It was carved out of tuff, the walls plastered and frescoes applied to the entire room. Then funeral goods, including items required in the next world could be placed by the urn or funeral bed of the deceased. There was joy and festivity associated with the passage to the next world.

THE ORIGINAL BLUE DEVIL

Yes, here he is, from an Etruscan tomb, Orcus I, mid IV B.C. a beak-nosed, wiry haired, bushy eyebrowed, bearded, snake holding, hammer wielding, azure wonder. This is Charun, escort to or keeper of the Underworld (a psychopomp). In Jungian psychology, these creatures are mediators between the unconscious and conscious realms. The blue skin may come from a reaction to snakebites or is symbolic of the decay of the flesh.

TOMBS: THE UNDERWORLD

PASSAGE TO HADES

ABOVE: Tomb of Orcus II, 325 B.C., fresco of Agamemnon, Tiresias and Ajax entering the Underworld through a marsh. Scholars feel this was the precursor for the model which Virgil used for Aeneas' and the Sibyl's entrance to the netherworld through the Stygian Marsh.

LEFT: A goddess of the Underworld, 450 B.C., sits on a throne guarded by two sphinxes. She holds a pomegranate, symbol of fertility and new life. This statue, hollow throughout, was placed in the tomb. The head can be removed, perhaps to pour libations to ancestors and the gods of the Underworld.

PASSAGE TO HADES
Chiusi 150 B.C., This cinerary urn (clay) shows a reclining man with an omphalos bowl which was thought to allow direct communication with the gods. The entrance to the Underworld is indicated by two arched doors between which two furies stand with torches. They threaten a woman with her child saying farewell to a priest. Hades is stepping out of a gate and next to him sits Charun. At the top is a wolf-headed demon.

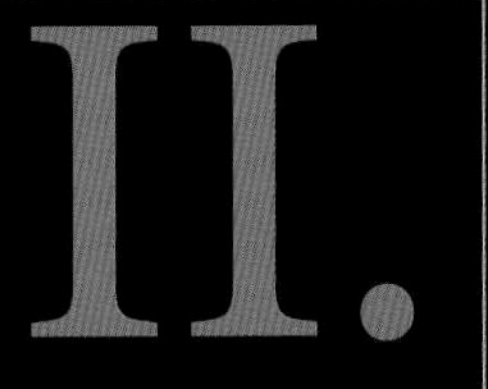

II. Hell on Earth: Into Virgil's Underworld

LAND OF INFERNAL ERUPTION
The Greeks called this place "the burning fields." To this day it emits boiling, poisonous vapors and mud. It is a manifestation of geothermal activity emanating from deep within the Earth. Virgil used it as a model for the Underworld.

Virgil (70-19 B.C.) chose to live in Naples for the same reasons as the ancient Greeks, its beauty and its mystery. He even had the audacity to refuse the request of the Emperor Augustus to live in Rome; preferring Naples. He saw it as the ideal setting for his Latin epic poem the *Aeneid*, written circa (29-19 B.C.). Because of the allure, mystery and fiery nature of the land, as well as its being the home of the legendary Sibyl of Cuma, Virgil found the perfect location for creating some of the most stirring passages in literature. The poem is divided into twelve books: the first six chronicle Aeneas' voyage to Italy from Troy after its fall in the Trojan War; the last six tell of him becoming the mythic founder of Rome around 753 B.C.

From a geologic standpoint, Book VI of this work is of particular interest for it describes Aeneas' landing on the shores of *Euboean Cumae* (VI.2), some 20 kilometers west of Naples and his eventual descent into the Underworld. While one might think that Virgil had a vivid imagination or based his ideas on the Christian concept of Hell (remember he wrote before Christ), in fact, his descriptions are based on real places, actual geologic occurrences and volcanic formations which exist today. Virgil only had to wander through the *Campi Flegrei* ("flaming fields") to find everything he would need to describe features of the Underworld; poisonous gases, bubbling mud, mysterious caverns, hot vapors and last but not least, an underground river. These geologic marvels, combined with temples to divinities and grottoes for the Sibyl, led him to create a saga that is spellbinding even today.

Founded by settlers from the Greek island of Euboea in the VIII century B.C., Cumae (Cuma in Italian) was the first Greek colony on the Italian mainland in what was once known as *Magna Grecia.* The Greeks may have chosen this location because of its high acropolis, a lava dome that would have allowed for adequate defense, or perhaps they were drawn to the beautiful yet volatile volcanic landscape, seeing in its flames and sulfurous vapors places celebrated in the poems of Homer.

Today, the Underworld Virgil describes in Book VI is still visible on the landscape of the Campi Flegrei, with elements of at least a half dozen of his Underworld features readily recognizable—among them the Grotto of the Cumaean Sibyl mentioned in verse 42; the Stygian Marsh (VI.133); Avernus, the Birdless Place (VI.237); the seething whirlpools and belching sands of the River Acheron (VI.295); the land of Shadows of the River Styx (VI.384); and the rushing flood of torrent flames—the Tartarean Phlegethon (VI.548).

A vast underground geothermal zone that stretches from Naples

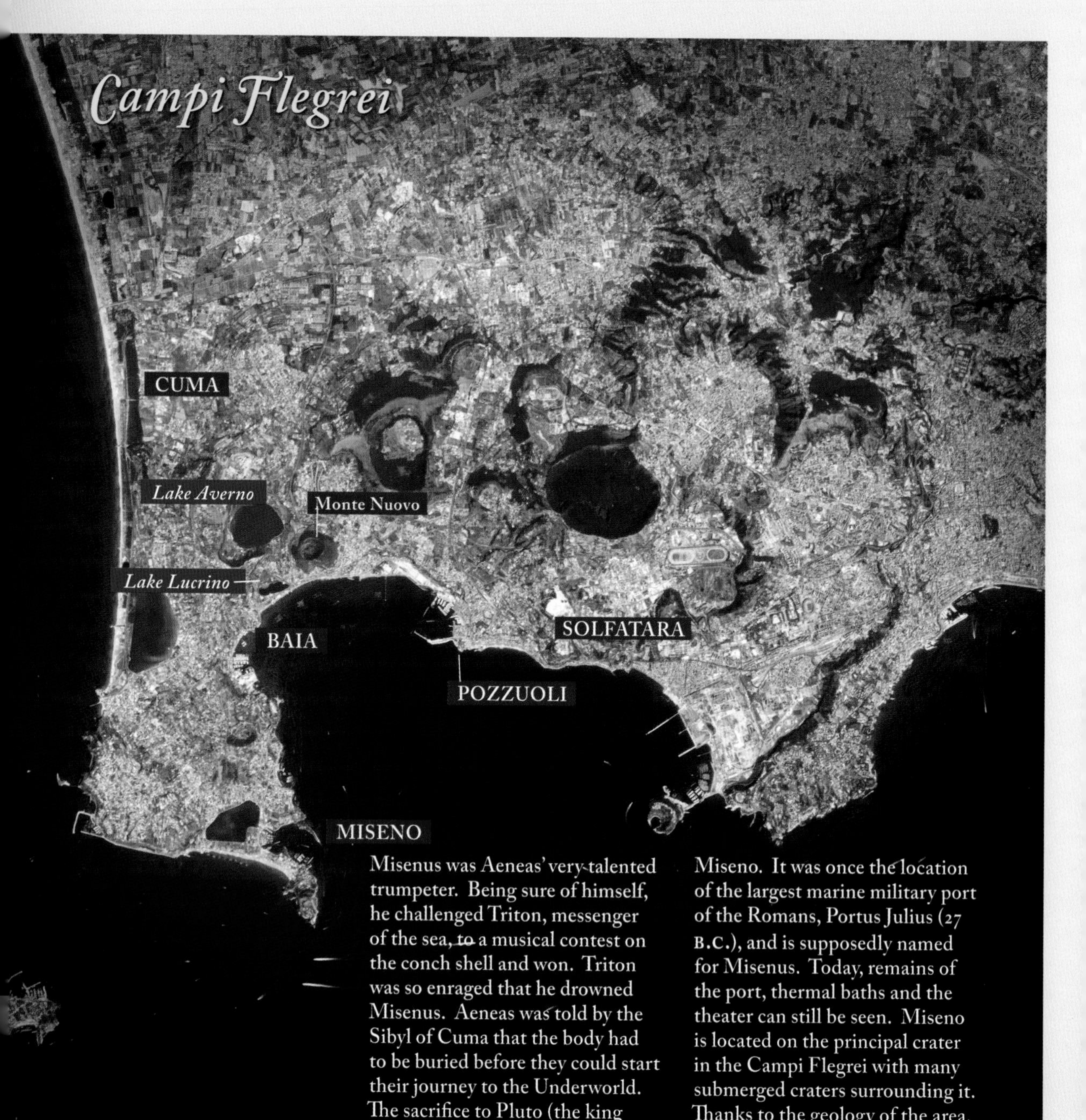

Misenus was Aeneas' very talented trumpeter. Being sure of himself, he challenged Triton, messenger of the sea, to a musical contest on the conch shell and won. Triton was so enraged that he drowned Misenus. Aeneas was told by the Sibyl of Cuma that the body had to be buried before they could start their journey to the Underworld. The sacrifice to Pluto (the king of the Styx) and to the infernal divinities started during the night in an area known today as Miseno. It was once the location of the largest marine military port of the Romans, Portus Julius (27 B.C.), and is supposedly named for Misenus. Today, remains of the port, thermal baths and the theater can still be seen. Miseno is located on the principal crater in the Campi Flegrei with many submerged craters surrounding it. Thanks to the geology of the area, the curvilinear configuration of the craters allowed the Roman fleet to be protected when anchored.

to Cuma, the Campi Flegrei, also known as the Phlegraean Fields (from *phlegos*, the Greek word for burning), is one of the most complex volcanic configurations in the world. The area is composed of 24 craters, most of which are submerged. The main caldera is 13 kilometers across and is thought to have been formed circa 39,000 years ago during an eruption that produced a deposit of pumice and cinder referred to as the Campanian Ignimbrite. The caldera, which lies half onshore and half offshore, is centered on the town of Pozzuoli—its floor pockmarked with smaller calderas and cones from later eruptions which are clearly visible in satellite images. Around the Solfatara crater, mythical home of Vulcan, the Roman god of fire, the ground continues to emit steam and boiling, sulfurous mud. The area is also subject to bradyseismic activity, a slow upward and downward motion of the Earth's crust, which is readily seen in the temple of Serapis in Pozzuoli, which now lies below street level.

In describing Aeneas' landing at Cumae, Virgil mentions that... *round keels fringe the beach* (VI.4), no doubt a reference to the two

TEMPLE OF SERAPIS
This was originally a Roman food market (*macellum*) I–II century A.D. Throughout the ages it has been subject to bradyseism, the slow moving up and down of the ground. Note that the temple is below street level, yet the pillars are still upright.

AENEAS MAKES LANDFALL AT CUMA
Virgil describes the landscape Aeneas found when he arrived at Cuma; sandy beaches leading to a sacred mount. Today, the setting remains as in Virgil's day, two rows of sandy dunes and a lava dome of yellow Neapolitan tuff.

stretches of dunes that run parallel to the Cumean coast formed by erosion of the mainland. The inland dunes host the *Silva Gallinaria*, a dense Mediterranean forest composed largely of ilex trees (*Quercus ilex*), a variety of oak prized by the Romans for ship construction. The forest was likely the model for the *Grove of Trivia* (VI.13) dedicated to Diana.

Once ashore, Aeneas' crew sets off to *...seek the seeds of flame hidden in the veins of flint...* (VI.7). A cryptocrystalline form of quartz, flint generates a spark when two pieces are rubbed together, ultimately igniting a fire. Meanwhile, Aeneas choses to seek *the heights, where Apollo sits enthroned* (VI.9), clearly the Temple of Apollo, the ruins of which can be seen today atop Monte Cuma, a trachyte (light-colored igneous rock) lava dome capped with younger yellow tuff. According to legend, the temple, said to once have been graced with golden doors, was constructed by Dedalus in the V century B.C. as a token of gratitude to the god for allowing him to land safely at Cuma, having fled the trap of Minos with wings of wax.

In describing Aeneas' descent into the Grotto of the Cumaean Sibyl, Virgil wrote, [as he approaches]: *the huge side of the Euboean rock is hewn into a cavern, into which lead a hundred wide mouths, a hundred gateways, from which rush as many voices, the answers of the Sibyl* (VI.42).

A cave fitting the description and known as the *Antro della Sibilla* (Grotto of the Sibyl) was identified by the renowned archeologist Amedeo Maiuri in 1932. It consists of an hexagonal gallery that is 131 meters long, terminating in an inner chamber carved out of tuff where it is believed the Sibyl was consulted on matters of great military, religious or civic importance. Her prognostications would be written on leaves that were ultimately dispersed by the wind.

Today, the tunnel, hewn out of yellow Neapolitan tuff by the Greeks in the IV-III century B.C., is a marvel to behold. Slivers of light stream in from strategically carved crevices allowing each hexagonal arch to be eerily illuminated by the light that changes color and intensity throughout the day. Neapolitan tuff is soft enough to allow a gallery of this extent to be carved with relative ease, yet

TEMPLE OF APOLLO
Aeneas would climb to the top of the lava dome which formed the acropolis to worship at the Temple of Apollo.

hard enough to provide structural integrity even after two thousand years. Legend has it that the hexagonal form, while aesthetically stunning, is said to have the power to deflect magnetic fields or negative energy that might enter the sanctuary.

In describing the approach to the entrance to the Underworld at Lake Averno, Virgil writes:

...Sprung from blood of gods, son of Trojan Anchises, easy is the descent to Avernus: night and day the door of gloomy Dis [Underworld] *stands open; but to recall one's steps and pass out to the upper air, this is the task, this is the toil...* (VI. 125-128).

...A deep cave there was, yawning wide and vast, of jagged rock, and sheltered by dark lake and woodland gloom, over which no flying creatures could safely wing their way; such a vapor from those black jaws was wafted to the vaulted sky whence the Greeks spoke of Avernus, the Birdless Place... (VI. 237-242).

Lake Averno derives its name from the Greek *aornos* meaning "lacking birds." In fact, the ancients wrote that fumes emanating from the lake were so poisonous that birds flying over it would die. We now know that the lake emitted lethal quantities of gases such as carbon dioxide (CO_2), hydrogen sulfide (H_2S), methane (CH_4) and sulfate (SO_4). The reason that these gases are present is that Lake Averno is crater lake, located over the vent of a volcano. The perimeter of its crater measures nearly two kilometers and encompasses an area of 370 acres with an incline that rises from 1 to 99 meters above sea level. The lake itself reaches 60 meters deep.

Ancient naturalists used the word "mephite" for certain lakes, grottoes and other places that infect the air with poisonous steam or vapor. In Roman mythology, Mefitis was the personification of the poisonous gases emitted from the ground in swamps and volcanic areas. Mephites actually do exist, with the *Grotta del Cane* (Grotto of the Dog) near Lake Averno being a famous example. There, Pliny the Elder (23-79 A.D.) observed that when a man

THE SIBYL OF CUMA

This hexagonal gallery was long thought to be the Sibyl of Cuma's cavern. It was carved by the Greeks out of yellow Neapolitan tuff. Legend has it that its form was designed to protect those inside from outside negative energies.

CVMAEA

The Sibyls (also known as oracles or pythia) were virgin priestesses with the gift of prophecy. They were found in areas which hosted cults of Apollo with the three main centers being Delphi, Eritrea (Turkey) and Cuma. The origin of the name Sibyl is not exactly known, but may come from the Greek, meaning "advice or prophetess from God." While there is agreement that the Sibyls actually did exist, we do not know their exact number, perhaps ten or twelve. Traditionally, their sanctuaries were located near a natural feature; a spring, mountain peak, crevasse, cave or even a place struck by lightning. Strabo (46-120 A.D.) wrote about *pneuma*, the gas or vapor emitted from a fissure in the earth. Geologic studies carried out by archeologist John Hale and geologist Jelle de Boer beneath the grotto at Delphi revealed an underlying layer of carboniferous limestone which emitted gases such as methane and ethylene which entered the cavern through faults. While the geologic configuration at Cuma is different (volcanic), the same gases found at Delphi are present. Such gases, presumably inhaled by the oracles, produced a trance in which their voices became low and raspy as they uttered their prognostications. Modern research shows that the effects induced by inhaling certain gaseous vapors are much akin to the high associated with sniffing model airplane glue.

The Sibyl of Cuma was famous in large part thanks to Virgil. No one knew how old she was, but according to legend, she held grains of sand in her palm that represented her age; the number of years which she requested Apollo grant her. He did so on the condition she leave her home in Eritrea (Turkey). While her wish for longevity was granted, she forgot to ask for eternal youth. At Cuma, she prophesied for many centuries but having lived so long, she only had the desire to die. According to legend, when the Eritresians sent her a missive sealed with clay, contact with earth from her native land negated Apollo's gift of longevity and she could finally die. Her remains were said to be kept in a stone vase in the Temple of Apollo at Cuma

According to some scholars, Virgil's *Eclogues* contain a Messanic prophecy by the Sibyl of Cuma, for this reason, in the Middle Ages, both the Sibyl and Virgil were considered prophets of the birth of Christ. Dante chose Virgil as his guide into the Underworld, and Michelangelo chose to place the Sibyl of Cuma as well as those of Delphi, Eritrea, Libya and Persia among the prophets of the Old Testament on the ceiling of the Sistine Chapel.

LAKE AVERNO
The Greeks referred to it as "the birdless place" because noxious vapors spewed from the lake, which is a volcanic crater. Virgil set the entrance to the Underworld on its shores. It has always been the source of fascination—from the XVI century French etching showing curious travelers—to today.

entered the cave with a dog (or other small animal) at his side, the animal would be asphyxiated while the man remained unharmed.

This unexplained phenomenon was a source of marvel and entertainment for visitors in the ensuing centuries, including those on the Grand Tour, such as Johann Wolfgang von Goethe, Alexandre Dumas and Sir William Hamilton (who as a scientist tried desperately to explain this mystery). Sometimes, the guide, wishing to add a new element to the routine, timed the animal's exposure so that it collapsed and appeared dead. He would then toss the "lifeless" body into the adjacent Lake Agnano where it would be reanimated. Visitors would be astounded by the power of the lake's "miraculous" waters.

It would be several centuries before scientists would discover that the cavern emitted carbonic acid (H_2CO_3), which being heavier than air, accumulates close to the ground. Carbonic acid is an unstable compound which readily decomposes into carbon dioxide (CO_2) and water (H_2O). When a man entered the cavern, being tall, he was not exposed to the poisonous vapors, but a small animal, walking close to the ground would inhale them and be asphyxiated.

As Aeneas nears the entrance to the Underworld, Virgil writes: *...just before the rays and dawning of the early sun the ground rumbled underfoot, the wooded ridges began to quiver, and through the gloom dogs seemed to howl as the goddess Hecate* [goddess of the netherworld] *drew* *nigh, "Away, away! You that are uninitiated!" shrieks the seer. "Withdraw from all the grove! And you, rush on the road and unsheathe your sword! Now Aeneas, is the hour for courage, now for a dauntless heart!" So much she said, and plunged madly into the opened cave; he with fearless steps, keeps pace with his advancing guide* (VI. 255-262).

Earthquakes such as the one just described were frequent in this area during what geologists call the third Phlegraean Period (8,000–500 years ago).

As for the entrance to the Underworld, Virgil tells us it is located on the shores of Lake Averno. To reach it, however, Aeneas must cross the river Styx, the river of hate, which marked the boundary between the living and the dead.

There is indeed a tunnel near the lakeshore, sometimes referred to as the Grotto of the Sibilla, which is at the terminus of a canal commissioned by Agrippa circa 37 B.C. and designed by the architect Lucius Cocceius Auctus. Its purpose was to link Lake Averno with Lake Lucrino and later the sea, transforming Lake Averno into Port Julius, home to the Roman fleet. Built as a secure pathway for

GROTTO OF THE DOG

For millennia, this mysterious grotto fascinated visitors as dogs or small animals walking into it were asphyxiated. The secret of this phenomenon would only be discovered in the last century, toxic gases which accumulated close to the ground.

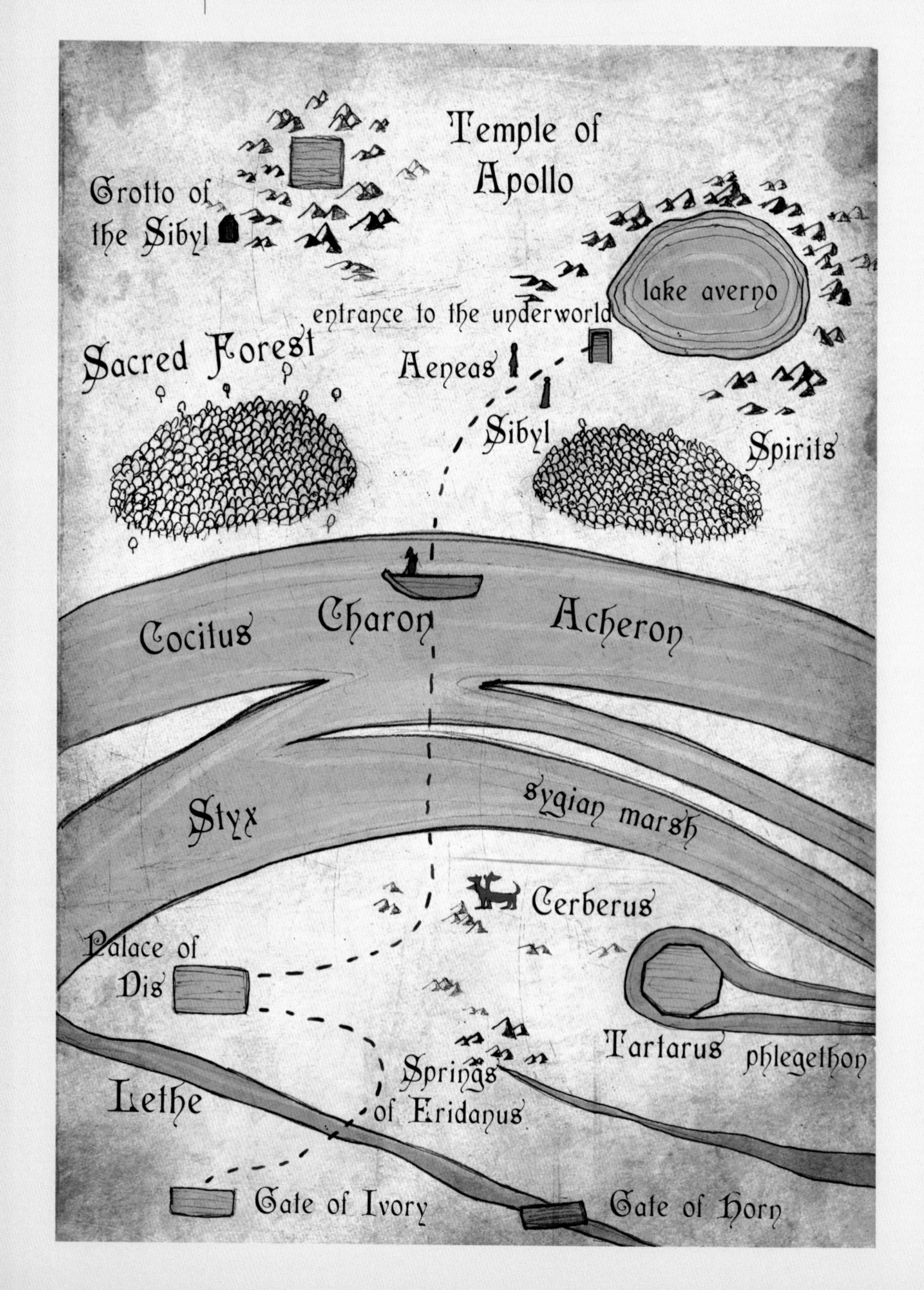
Temple of
Apollo
Grotto of
the Sibyl
lake averno
entrance to the underworld
Sacred Forest
Aeneas
Sibyl
Spirits
Charon
Cocitus
Acheron
Styx
sygian marsh
Cerberus
Palace of
Dis
Tartarus
phlegethon
Springs
of Eridanus
Lethe
Gate of Ivory
Gate of Horn

Roman legions, the tunnel itself is some 200 meters long and about 4 meters high. Having been damaged in the earthquake of Monte Nuovo in 1538, it is not often frequented. However, an expedition into the depths of the foreboding tunnel reveals a set of anterooms which are filled with water. Legend holds that they were the ritual baths used by the Sibyl for spiritual purification. This type of space has always been associated with a cult of the dead, leading to the belief that this is where the Sibyl allegedly came to die.

Moving deeper into the tunnel is a small side corridor with a tiny stairway which descends about five meters. There, in the glimmer of torchlight, one can see an extraordinary thing—an underground river which flows mysteriously down, deep into the Earth. It is unclear whether this river is a remnant of an aqueduct that may have been part of the Roman era military construction or is a natural water-bearing subterranean fault or seam formed by bradyseism. The underground river is likely the model Virgil used for the river Styx as it is in the exact location, right next to Lake Averno in a deep, dark cavern.

ENTRANCE TO THE UNDERWORLD
This tunnel on the shore of Lake Averno was excavated by the Romans. Virgil used it as the entrance to the Underworld. It contains an underground river.

According to Greek mythology, upon which Virgil drew heavily, there were four Underworld rivers in addition to the Styx—the Cocytus, the river of lamentation; the Acheron, the river of pain and woe across which the newly dead are ferried and from which the Styx and Cocytus spring; the Lethe, the river of forgetfulness; and the Phlegathon, the molten river of flames—all of which converge in the Stygian Marsh.

Ample wetlands surround Lake Lucrino and for geologic reasons leads us to believe it was the model for Virgil's Stygian Marsh.

THE STYGIAN MARSH
Lake Lucrino (foreground) was once navigable. Bradyseismic activity caused its waters to recede and it was soon filled with mud. In its new form as a swamp, Virgil used it as a model for the Stygian Marsh.

Between 31 and 12 B.C.—not long after the construction of Port Julius—an episode of bradyseismic activity led to a dramatic lowering of the water level in Lake Lucrino such that it was no longer navigable and ultimately became little more than a swamp. As a result, the port was moved to Miseno.

As Aeneas continues his journey he is frightened by the dark caverns he must pass.

The Amphyrsian soothsayer spoke briefly: "No such trickery is here; be not troubled; our weapons offer force; the huge doorkeeper may from his cave with endless howl affright the bloodless shades" (VI. 399-401).

Many believe these caverns are natural, but the majority are not. From the time of the Greek arrival at Cuma the terrain was excavated for building material. Two areas were created, one above ground, where the rock was used to construct cities and one below ground where, in the caves left void from the rock excavation, catacombs, aqueducts and defensive fortresses were created.

The ancients thought that the Cimmerians, a mythical

subterranean population excavated them. No one can give an exact definition of this populace because no archeological evidence has been found confirming its existence. The first written citation was in Homer, who referred to them in the *Odyssey.* Cicero placed them at Cuma and Strabo claimed they lived between Averno and Baia, describing them as a populace who inhabited subsurface caverns called *argille.* They reached one another's homes through a series of tunnels. Allegedly, they lived by selling the minerals they mined and the donations of visitors who came to consult their oracle. What little we know is that they might have been early Greeks who were a part of a cult of the night and were prohibited from seeing the light of the sun. They also could have been miners who spent the whole day below ground. Or, were these the mysterious spirits who frequented the shores of Lake Averno and whom the ancient writers identified as the dead who exited the Underworld?

Two of Virgil's later passages, which describe the Acheron and Phlegathon appear to have been inspired by a frightening area in the Campi Flegrei known as the Solfatara.

From here the road leads to the waters of Tartarean Acheron. Here

CAVE COUNTRY
The volcanic tuff was easily excavated, leading to an extensive network of tunnels and galleries. Legend has it that a mythical populace, the Cimmerians lived in perpetual darkness underground.

thick with mire and of fathomless flood, a whirlpool seethes and belches into Cocytus all of its sand... (VI. 295-297).

...Suddenly Aeneas looks back, and under a cliff on the left sees a broad castle, girt with triple wall and encircled with a rushing flood of torrent flames—Tartarean Phlegethon, that rolls along thundering rocks... (VI. 547-550).

Formed between 3,700 and 4,600 years ago, the Solfatara is a hydrothermally altered tuff cone (a large circular depression with a low rim of volcanic debris), which was then subjected a post-eruption caldera collapse. It is the most volcanically volatile area of the Campi Flegrei.

In the Solfatara caldera, there are active fumaroles, the largest of which is Bocca Grande. In the center is a bubbling pool with a fissure from which water and clay are emitted to combine into a mud consisting of water, carbon dioxide and sulfuric acid, with temperatures ranging from about 170° to 250°C. The muddy mixture also contains trace amounts of barium, sodium, magnesium, vanadium, arsenic, zinc, antimony, iodine and rubidium. Surrounding deposits include colorful sulfur compounds such as realgar, cinnabar and orpiment.

Virgil would have seen the steamy vapors, bubbling mud and noxious gases emitted from a rumbling Earth, which in his day could not have been explained scientifically. One can imagine that he would have been astounded, mystified and frightened by the Campi Flegrei. Yet, despite the violent nature of the land and the gripping legends associated with it, he explored and found there, the inspiration to create one of the greatest literary works of all time.

THE FLAMING FIELDS

Virgil's models for the Underworld: ABOVE and BELOW LEFT: the Solfatara, where the Earth continues to boil, emitting vapors, mud and sulfurous minerals. BELOW RIGHT: the pockmarked landscape, with Lakes Averno and Lucrino in the distance.

Christ Stopped at Lake Averno

According to Christian belief, after dying on the Cross, Jesus Christ spent forty days on Earth, descending into Hell before ascending into Heaven. In medieval times, a strange legend came into being. It held that after His death, Christ stopped at Lake Averno, the entrance to the Underworld, to liberate those souls who had abided by the ancient laws. This is referenced in a poem by Pietro d'Eboli (1170?-1220?), *De balneis Terrae Laboris*:

> *Est locus Australis, quo porlam Christus Averni*
> *Fregit, et eduxit mortuos inde suos.*
>
> There is a place in the south, Averno, in which Christ
> broke the gate and lead forth his souls.

A geologic aspect was added to the legend. Jesus, exiting from the Underworld at Lake Averno, blocked the entrance with the Mountain of Christ the Savior. This mountain is visible today and is called Monte Gauro.

While very little is known about the life of Pietro d'Eboli, 35 short poems entitled *Baths of Puteoli* (modern day Pozzuoli) were written in 1197. They were illustrated in the XIII century and bound in a codex now in the Biblioteca Angelica of Rome. This miniature, *Balneum Tripergulae* illustrates the baths which were located near Lake Averno. The lower section shows Christ destroying the gate to Hell. The baths were renowned for their curative powers. Unfortunately, the village in which they were located was destroyed by the eruption of Monte Nuovo in 1538.

Christ Stopped at Lake Averno

UPPER LEFT: *De Balneis Puteolanis.* These baths are located today at Agnano. The miniature shows a sauna and a man intending to cool off a bather by drawing a pitcher of fresh water from the lake, infested by frogs and snakes. At the top is San Germano who came to pray for the soul of deacon Pascasio, suffering in Purgatory amid burning vapors.

UPPER RIGHT: *Balneum Sulphatara.* In medieval times, this was considered a miraculous place for curing infertility in women. The miniature shows a group of women bathing in an hexagonal tub (symbolic of fecundity) while in the background a person fans the fumaroles to emit more vapors.

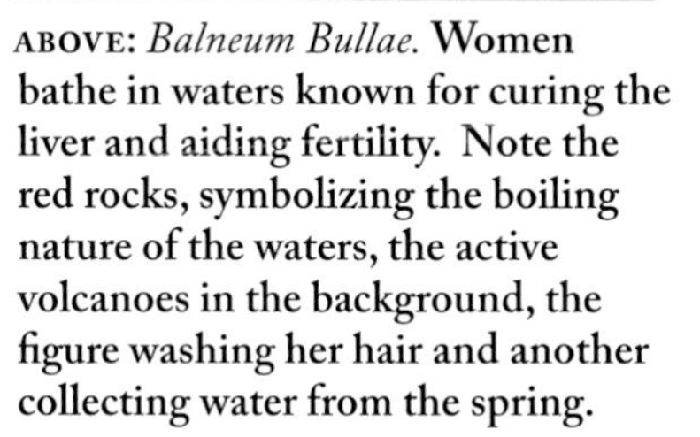

ABOVE: *Balneum Bullae.* Women bathe in waters known for curing the liver and aiding fertility. Note the red rocks, symbolizing the boiling nature of the waters, the active volcanoes in the background, the figure washing her hair and another collecting water from the spring.

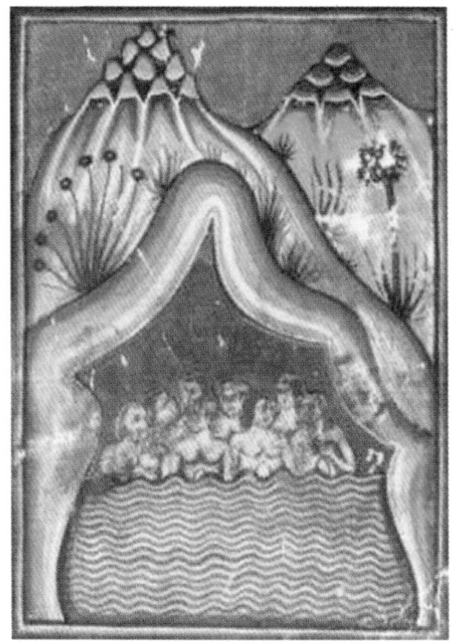

ABOVE: *Balneum Spelunca.* Near the Castle of Baia, these waters were used to heal gout and chronic coughs. Located near the ancient Roman baths, they are surrounded by volcanoes. The text notes that this cavern is not frequented by thieves or persons of ill repute.

Vedi Napoli e poi muori *
—Goethe

Pyroclastic Poets

Poets were not immune to the mysterious and terrifying allure which volcanoes offered. Throughout history, writers joined the ranks of explorers and scientists in venturing close—either actually or metaphorically to the fire spewing vents which flaunted their hot vapors, ruby colored flames and glowing briquettes like a circus juggler, enticing anyone with enough moxie to venture near.

Poets found volcanoes mysterious and fraught with hazards, yet at the same time, spellbinding and erotic. In them they found the ideal vehicle for metaphor, allegory and symbolism. Whether it was a commentary on politics as found in Shelley, a choice between God and the Devil as in Goethe, or the wistfulness of never having seen a volcano by Emily Dickinson.

Volcanoes have been used as the basis for myth since Classical times: Mt. Etna was symbolic of the one-eyed Cyclopes who

* See Naples and die.

menaced Ulysses in the *Odyssey*. The Greek poets of Sicily produced a plethora of works on the various eruptions. Interestingly, this extraordinary literary production captured the interest of scholars in other disciplines. Plato traveled there in the IV century B.C. and ascended Mt. Etna in an attempt to elaborate on the concept of central fire. Strabo, the geographer, in the I century B.C. described the eruptive phases of the volcano in his popular text *Geography*.

Eventually, the Greek fascination with the subject would pass to the Latin scholars such as Virgil who wrote:

How oft before our eyes did Etna deluge the fields of the Cyclopes with a torrent from her burst furnaces, hurling thereon balls of fire and molten rocks ...(Georgics, I. 468).

And so it continued, with the volcanoes capturing the imagination of man throughout the ages.

Johann Wolfgang von Goethe

The German poet arrived in Naples in 1787 and was hosted by Sir William Hamilton, who in addition to being the British envoy to the Court of Naples, was a scientist who wrote extensively on Vesuvius. For Goethe, the site of the volcano erupting was overwhelming. Despite his fears, including one of heights, he climbed the smoking mountain three times. No scene would ever provide him with a combination so thrilling yet so calm *"the emotions and the senses (were) in complete equilibrium"* he wrote.

Upon touring the ruins of Pompeii in the 1780s, some 40 years after its rediscovery, he wrote, *"Many disasters have befallen the world, but few have brought posterity so much joy."*

After surviving a burst of hot stone and ash which shot out of the crater while standing on the rim, the poet understood the Neapolitan psyche, commenting *"The Neapolitan would certainly be a different creature if he did not feel himself wedged between God and the Devil."*

The struggle between the forces of God and the Devil would become a central theme in Goethe's masterwork, *Faust*, in which he describes the satanic Mephistopheles in

IN-DEPTH

Pyroclastic Poets

ICONIC MT. ETNA **This stately symbol of Sicily, a stratovolcano, has been erupting continuously for 3,000 years.**

Vesuvian metaphors:

The Devil's all set up a coughing, sneezing,
At every vent without cessation wheezing;
With sulphur-stench and acids Hell dilated,
And such enormous gas was thence created,
That very soon Earth's level, far extended,
Thick as it was, was heaved, and split,
and rended!

Like Vesuvius, Sicily's Mt. Etna had its hold on Goethe, who wrote in his book, *Italian Journey: 1786–1788*:

...walks on foot through the most astonishing landscape in the world; treacherous ground under a pure sky; ruins of unimaginable luxury, abominable and sad; seething waters; caves exhaling sulfur fumes; slag hills forbidding all living growth; barren and repulsive areas; but then, luxuriant vegetation, taking root wherever it can, soars up out of all the dead matter, encircles lakes and brooks, and extends its conquest even to the walls of an old crater by establishing there a forest of noble oaks.

It was in fact, Goethe's writings which helped to promote Mt. Etna and Vesuvius as a "must see" on the Grand Tour in which young Britons visited the continent to complete their education.

Percy Bysshe Shelley When the poet came to Naples in 1818 with his wife Mary, he hired guides to take the couple to the summit of Vesuvius, where, upon seeing a river of hardened lava, he wrote *"an actual image of the waves of the sea, changed into hard black stone by enchantment."* Upon reaching the summit, he peered into the crater and wrote, *"After the glaciers, the most impressive exhibition of the energies of nature I ever saw."*

VESUVIUS **After visiting the volcano in 1867, Mark Twain commented** ***"One could stand and look down upon it for a week without getting tired of it."***

The intensity of the experience brought him close to a breakdown and he wove his impressions of the volcano into his play, *Prometheus Unbound.*

Shelly returned to Naples the following year, and upon traveling to Pompeii, wrote:

Ode to Naples

I stood within the city disinterred,
And heard the autumnal leaves
like light footfalls
Of spirits passing though the streets,
and heard
The Mountain's slumberous voice at intervals
Thrill the roofless halls:
The oracular thunder penetrating shook
The listening soul in my suspended blood;
I felt the Earth out of her deep heart spoke...

Mary Wollstonecraft Shelley wife of the poet and author of *Frankenstein,* was also fascinated with the mysterious volcanic region. In her novel, Dr. Victor Frankenstein states that he was born in Naples. In her 1826 work, *The Last Man,* she describes a frightful trip into the Grotto of the Sibyl of Cuma. The description is based on her visit to the cavern during her 1818 stay in Naples.

Emily Dickinson While never having seen volcanoes in Massachusetts, the poet described them with great emotion in a poem included in her *Master Letter 2* (1858-1862):

IN-DEPTH

Pyroclastic Poets

I have never seen 'Volcanoes'—
But, when Travellers tell
How those old—phlegmatic mountains
Usually so still—

Bear within—appalling Ordnance,
Fire, and smoke, and gun,
Taking Villages for breakfast,
And appalling Men—

If the stillness is Volcanic
In the human face
When upon a pain Titanic
Features keep their place—

If at length the smouldering anguish
Will not overcome—
And the palpitating Vineyard
In the dust, be thrown?

If some loving Antiquary,
On Resumption Morn,
Will not cry with joy 'Pompeii'!
To the Hills return!

Charles Dickens arrived in 1844. He described it as *"The mountain is the genius of the scene."* Dickens and his party climbed Vesuvius the day he arrived, despite the fact it was covered in snow. At the top, Dickens writes *"There is something in the fire and roar that generates an irresistible desire to get nearer to it."* When he did, he stared into *"the hell of boiling fire below."*

Alexandre Dumas The French writer was drawn to Italy and the story of Vesuvius by chance. In the 1840's Dumas had written the highly popular *The Three Musketeers* and *The Count of Monte Cristo.* But by 1860, at fifty, he was broke and his work ignored. To add insult to injury, his son and namesake, Alexandre Dumas, *fils*, had written the play *Camille* and was very successful economically and professionally. The elder Dumas decided he needed a change and happened to meet Giuseppe Garibaldi during the struggle for Italian unification. Dumas, using his own boat, became a gun runner for Garibaldi. At the end of the war, Garibaldi appointed him head of the excavations at Pompeii and Herculaneum. While he only lasted a few years on the job, it was enough for him to catalog the ribald art which is now on display at the National Archeological Museum in Naples.

Mark Twain Not one to miss some excitement, Mark Twain arrived in 1867 and found Vesuvius and Pompeii deeply moving. He describes his travels in the *Innocents Abroad*:

"The crater itself—the ditch—was not so variegated in coloring, but yet, in its softness, richness and unpretentious elegance, it was more charming, more fascinating to the eye. There was nothing "loud" about its well-bred and well-creased look. Beautiful? One could stand and look down upon it for a week without getting tired of it."

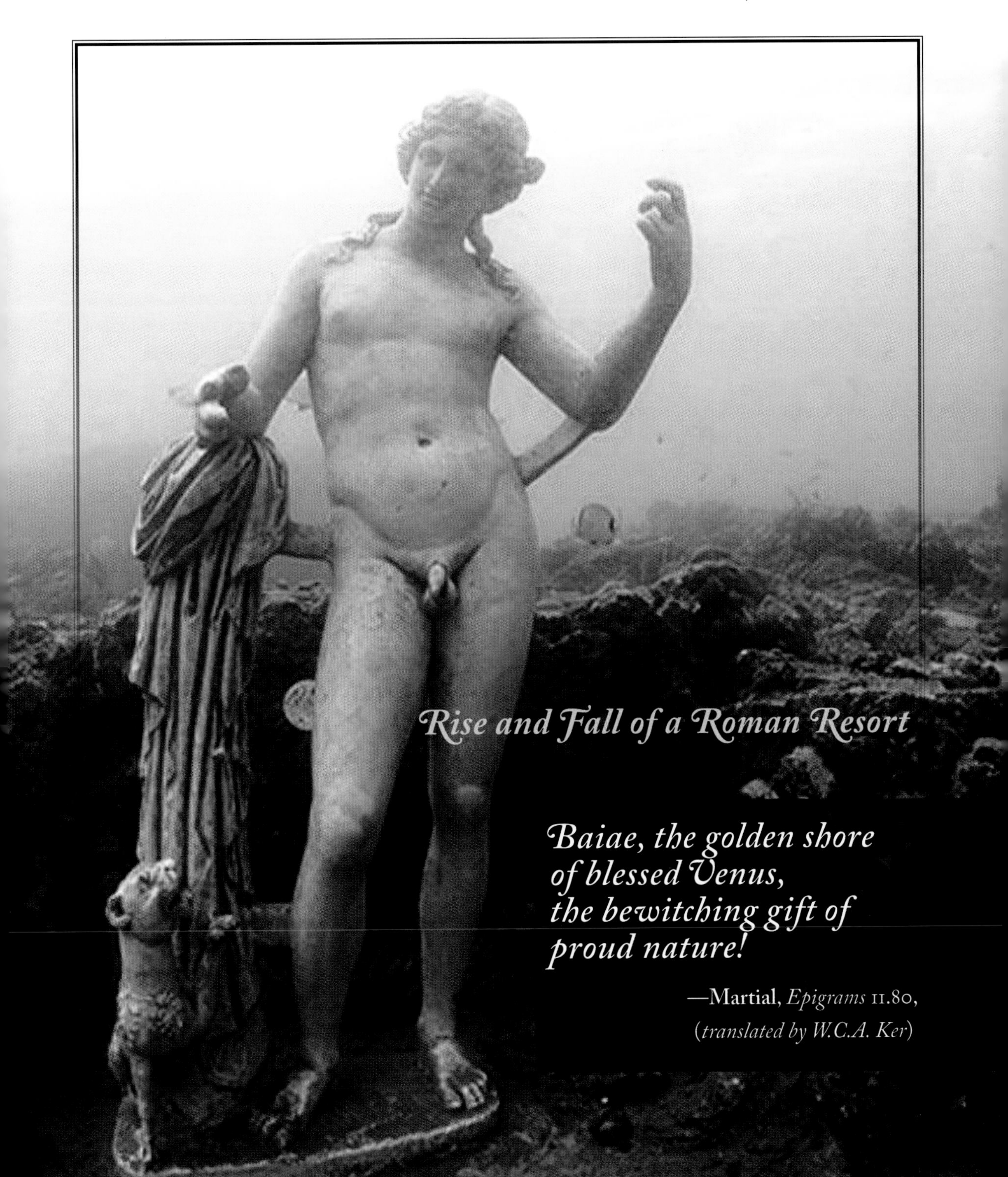

Rise and Fall of a Roman Resort

> *Baiae, the golden shore*
> *of blessed Venus,*
> *the bewitching gift of*
> *proud nature!*
>
> —Martial, *Epigrams* 11.80,
> (*translated by W.C.A. Ker*)

IN-DEPTH

Rise and Fall of a Roman Resort

Like modern-day Monte Carlo or St. Tropez, Baia (Baiae in Latin) was known for its high rollers—emperors Julius Caesar, Nero and Hadrian among them—who frequented the resort, living life large in opulent surroundings. A number of Roman authors described in detail the luxuriant town and the activities that took place there. Adding to the excitement was the fact that it was close to the newly built Port Julius.

And thus, Baia was like the locus of a perfect storm: people, power, plenty and a bit of perversion thrown in. Wild parties and even wilder demonstrations of wealth were everywhere. On the bay, large ships served as theatrical stages, hosting actors, animals and extraordinary fireworks to celebrate just about anything. On land, the earth bubbled and boiled away providing hot healing muds, therapeutic baths and ethereal vapors which seeped out from the underworld day and night. As if this wasn't enough, the land decided to aid these perpetual party people by turning itself into a stage—one that went up and down.

If one travels to Baia today, its beauty is breathtaking: the perfect little bay with Vesuvius lounging in the background and Capri and Ischia off in the distance. Set back from the shore is a vast complex which was once the thermal spa for some of the most powerful people in history. The thermal waters were abundant in this volcanic area, the Campi Flegrei, and the Roman engineers designed sophisticated piping to bring in the water, steaming hot. They did this by tapping into underground scalding liquids and channeling them to the spa where they were used in the saunas. The water contained natural salts and minerals which were efficacious for treating any number of illnesses. The Roman doctors on the staff of the spa were the best in the

BAIA BELOW SEA LEVEL
A nymphaeum (first half of the I century A.D.) with exquisite statues was discovered in 1969. It is now thought to have been constructed by the Emperor Claudius. In the niches, four statues were found: Antonia Minore, mother of the emperor Claudius, a figure of a child identified as Ottavia, the daughter of the emperor who died as an infant, and two sculptures of Dionysis, god of wine. The Homeric influence was represented by the intoxication of Polyphemus, with Odysseus offering a wine goblet to the Cyclops (the Cyclops is inferred, as only a lock of sculpted hair was found, presumed remnants of the configuration of this scene). The architecture of the room and the presence of hydraulic piping to control the water flowing into it provide proof of its use as a nymphaeum. The statues were recovered from their underwater resting place and brought into the Archeological Museum at Baia, where a nymphaeum was recreated.

IN-DEPTH

Rise and Fall of a Roman Resort

THE MOST LUXURIOUS SPA IN THE ROMAN WORLD
Frequented by the most powerful individuals in the empire, the baths were opulent.

empire and they were responsible for the welfare of the ruling class.

Today, one can walk on the mosaic floors which covered its steam rooms and saunas and see traces of the frescoes in the small rooms where the Roman elite were massaged. The complex continues on various levels, in a state of such magnificent preservation that one can actually imagine the whispers and laughter of 2,000 years ago. The spa's vast marble terraces overlooking the water were designed to allow privacy while relaxing, yet the ability to look down at the bay and of course the exquisite villas that once dotted the shoreline. Yet looking down from the terrace today, there is not a trace of a villa. A road winds along the coast and restaurants offer the catch of the day. Those villas did not share the good fortune of the thermal spa. There is nothing left of them.

But if you go down to one of those restaurants for lunch, maybe someone will tell you this amazing story. More than 50 years ago, veteran Italian fighter pilot Raimondo Bucher first spied the 2,000-year-old submerged remains of Baia. Taking to the sky on a clear day in 1956, Bucher banked the plane over the Gulf of Pozzuoli as he left the departure pattern of Naples' Capodichino Airport. It was then, as the crystalline waters slipped beneath his wing tip that he caught sight of what appeared to be walls and streets, and the outlines of buildings and harbor facilities.

Rise and Fall of a Roman Resort | IN-DEPTH

His curiosity piqued, Bucher, who was a skilled diver, decided to return to the site to explore first-hand what he had seen from the air, coming upon "mosaic floors of indescribable beauty and imposing columns."

How these abundant archaeological remains have come to be both underwater and well preserved is due to a mysterious maneuver of the earth called bradyseism. This movement, in which the land rises and falls, is caused by the fluctuating volume of a magma chamber below the surface or, as a result of hydrothermal activity. Bradyseism—the word for which comes from the Greek "bradus" meaning slow and "sism" meaning movement—can be likened to a theatrical stage that moves vertically, up and down, without toppling objects upon it. And that is exactly what happened to the villas which once dotted the coast. Over the past 2,000 years they have been sporadically raised and lowered as much as nine meters below sea level. There is no other place in the Mediterranean where the subsidence has been as dramatic as at Baia. For volcanologists, Baia is a godsend as they rarely encounter an area where ground movement can be measured so accurately. And the measurements are not only based

THE SPA AT BAIA
The thermal complex was built on many levels, affording stunning sea views and many nooks and crannies for therapy and privacy. The spa offered saunas, steam rooms, hot baths, massages as well as medical consultations. It was the exclusive enclave of the Roman elite.

on submarine lithic deposits, but by the actual dates on inscriptions, monuments and historical records.

After a half century of exploration of this one-time pleasure capital, archaeologists have identified the warehouses, docks and storage facilities associated with the original Port Julius—part of an ambitious civil engineering project commissioned by Augustus in 37 B.C. to link Lake Lucrino to Lake Averno. Also underwater was a magnificent nymphaeum attributed to the emperor Claudius which was adorned with stupendous statues and palatial villas with swimming pools, private docks and special ponds in which prized moray eels were raised. Baia, however, has yet to reveal all of its secrets. Only recently, an amphitheater was identified in the waters below the Castello Aragonese Museum which now houses Claudius' nymphaeum and archaeological finds from the Campi Flegrei.

Today, the submerged remains of Baia lie within a protected underwater archaeological park. One can still see the most extraordinary remnants of the sumptuous life enjoyed by Rome's rich and powerful barely 15 feet below the surface, or at arm's length, if one is inclined to scuba dive. All of these archeological marvels are set off not only by the clear blue waters, but by the effervescent bubbles produced by fumaroles on the sea floor.

Who knows, under the principles of bradyseism, the earth can rise as well as fall. Maybe one day, the villas will rise up from their resting place and a new party can begin.

III. Paradise Bejeweled

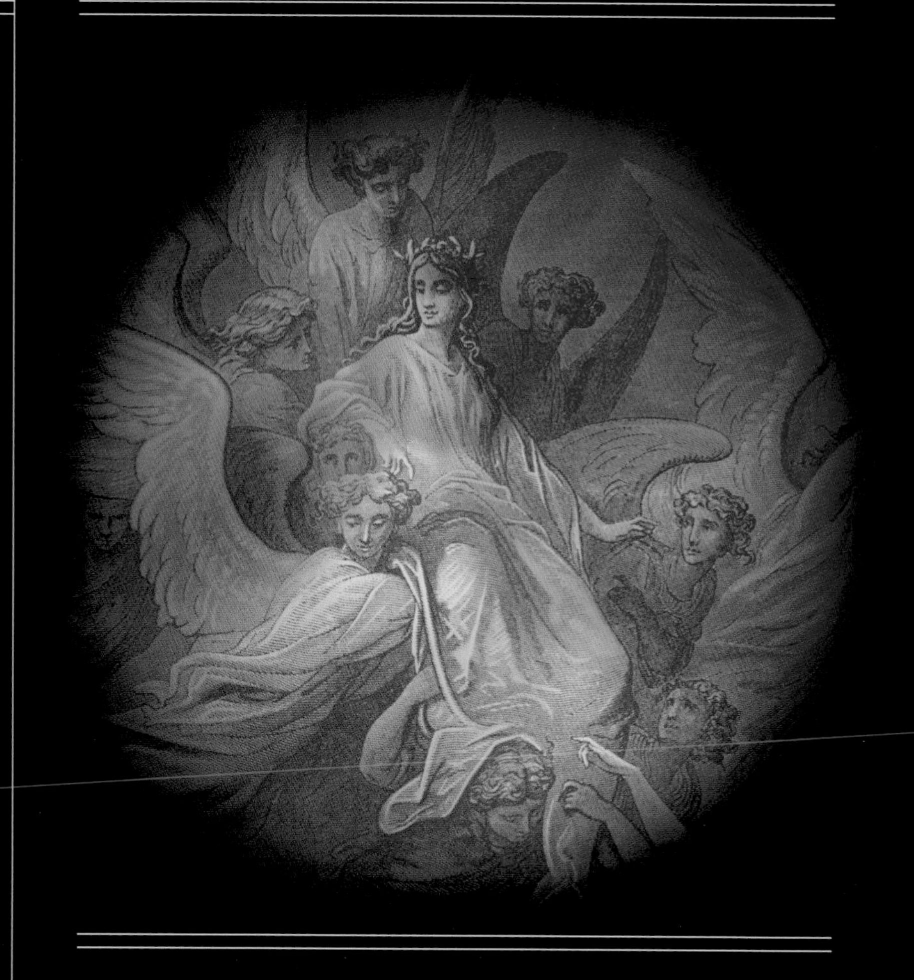

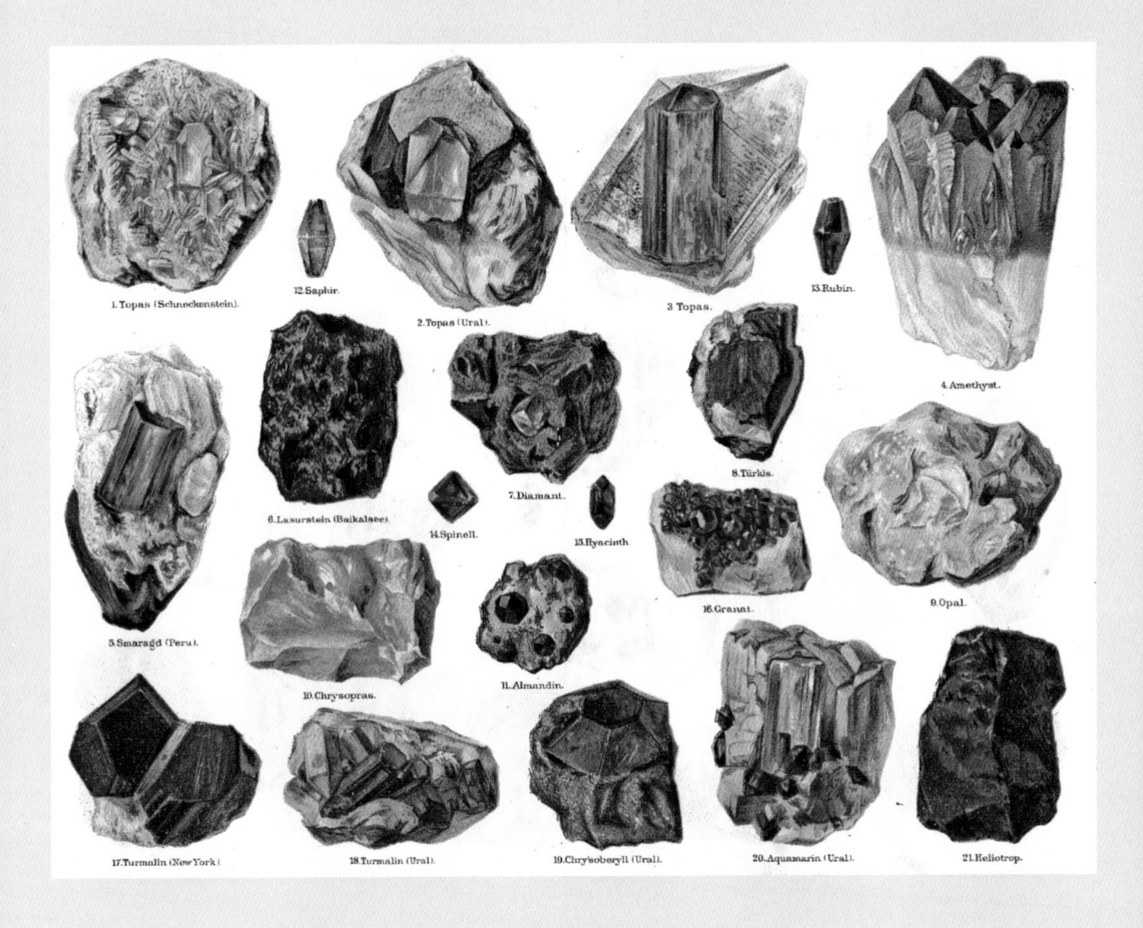

Ne la corte del cielo, ond' io rivegno,
si trovan molte gioie care e belle
tanto che non si posson trar del regno;

Within the court of Heaven, whence I return,
Are many jewels found, so fair and precious
They cannot be transported from the realm;

(Paradiso, X, 70–72)

Painted in 1465, nearly 150 years after Dante's death, Domenico di Michelino depicts the poet, poem in hand, in the center of Florence surrounded by the souls in Purgatory making their way to the heavenly realm. In the background is the recently completed dome of Santa Maria del Fiore, designed and constructed by Filippo Brunelleschi.

Hailed as the father of the Italian language, Dante Alighieri (1265–1321) is best known for his *Divine Comedy (Divina Commedia)*, penned between 1308 and 1321. The work was revolutionary in that it was the first book to be written in Italian, known as the Florentine *Volgare* rather than Latin, which was the language of the elite. Although inspired by Homer's great epic poems and Virgil's *Aeneid*, Dante's *Divine Comedy* chronicles a personal journey—his spiritual pilgrimage through the realms of Hell (*Inferno*), Purgatory (*Purgatorio*), and Paradise (*Paradiso*) in search of the ultimate truth revealed by God.

In the seven centuries since its composition, the *Divine Comedy* remains one of the greatest literary accomplishments of all time and as such, has received no shortage of scholarly analysis and commentary. Translations of the work tend to be divided between those faithful to the poet's words and those in which certain liberties have been taken in an effort to preserve its poetic structure regardless of the language. Several translators have also attempted to retain the *terza rima* (aba, bcb, cdc…) structure of the cantos for which the *Divine Comedy* is famed. (For the purposes of this discussion, we will be drawing on the Italian poetry with Henry Wadswoth Longfellow's English translation.)

Beyond the literary merits of the *Divine Comedy* the triptych overflows with insightful observations on the political and ecclesiastical upheaval that engulfed Dante's native Florence in the Middle Ages. It presaged the nascent intellectual rumblings that would later become the Humanist movement and influenced the works of Francesco Petrarca, "Petrarch" (1304–1374), Giovanni Boccaccio (1313–1375) and Leonardo Bruni (ca. 1370–1444). Dante's mastery of numerology has also been long appreciated by scholars. Throughout the poem, he makes numerous references to twelves; the apostles, sevens; the deadly sins, and threes; the Trinity.

Even today, the work has also attracted scientists looking for insight into the state of "scientific knowledge" in the early 1300s, particularly the poet's comments on physics and astronomy. Dante understood the spherical nature of the Earth, the placement of land masses upon it, the differing stars in the southern hemisphere and the nature of meridians. While he believed, as did Aristotle, that the Earth was the center of the universe, he attributed the movement of the nested spheres of heavenly bodies to a divine system of celestial mechanics.

His study of time keeping devices—which were undergoing a revolutionary change from water-driven to mechanical, led to his commentary on the nature of time zones. In the ensuing centuries, Florence would become the center of map making, allowing for the known world to expand exponentially in the Age of Exploration.

With all this scholarly inquiry however, there remains yet another area worthy of note. That is Dante's vast knowledge of gems and the emerging field of gemology. From a geological point

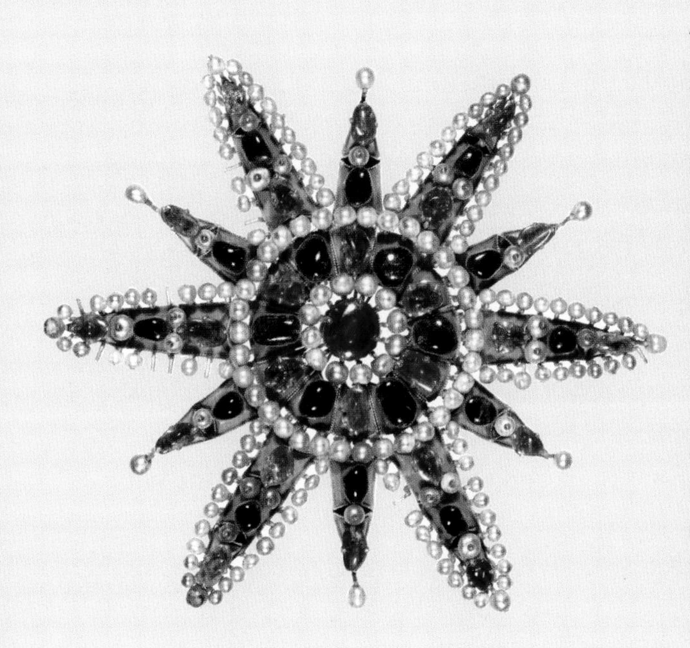

This stupendous 1335 brooch was made in Venice and is replete with numerological and astronomical significance. The center jewel, signifying the divine is circled by 12 pearls and 12 gems. Radiating out from the center on the six longer axes are 8 stones, which according to jeweler and historian Alberto Zucchetta, correspond to the planets: the Earth, Moon, Mercury, Venus, Sun, Mars, Jupiter and Saturn, a jeweled version of Dante's nested spheres.

of view, the *Divine Comedy* is a veritable treasure trove, containing pearls, rubies, topazes, emeralds, sapphires and diamonds, as well as crystal, amber and glass. Most of these gemological references can be found in the *Paradiso*, the Canticle of Light, in which Dante makes abundant use of illumination on objects in the form of reflection, refraction and shadow to convey a variety of metaphors and concepts—pearls, the intellectual lustre of the wise; rubies, souls of Christian warriors; diamonds, fortitude and steadfastness; and the sapphire, emblematic of the Virgin Mary, Queen of Heaven.

It is evident that Dante was well aware of the intrinsic physical characteristics of each gemstone and its astrological association as well as the spiritual, metaphysical and medicinal attributes each was purported to possess. His working knowledge of light's reflection, refraction and dispersion on specific gems is extraordinary as he combines the thought process of a physicist with the words of a bard. All of this in an era in which many rare, faceted precious stones were entering Europe and the principles of gemology, as we know them, lay centuries in the future.

Gems in the Age of Dante

For millennia, jewels and precious metals have been used as talismans and amulets, acquiring rich symbolic associations. Yet the capacity of lapidaries to actually cut precious stones is a relatively recent development—the first faceted gems arriving in Europe from India and the Far East following the fall of Constantinople in 1204.

At that time, Italy, which had played a critical role as a staging area for the Crusades and their associated commercial activity, was becoming a primary entrepôt for the importation of products from the East, many of which were previously unknown or that had been rare and costly to acquire—silks and calicoes, spices and exotic produce, ivory, gunpowder, improved methods for glassmaking and precious stones. With its skilled cadre of goldsmiths, Italy quickly rose to preeminence as a center for stone setting. It was also at this time that Europe became reacquainted with scientific and literary works of Classical antiquity including texts on gems, minerals and natural history which had been out of circulation since the fall of Rome in the late V century A.D.

While polyhedral faceting had developed in the East and was used to shape rubies, sapphires, emeralds and other comparatively "soft" stones, most gemstones continued to be cut in cabochon, that is, gradually brought into a round and regular shape through a tedious process of bruting and buffing. Diamonds, the hardest of all materials, were valued for their rarity and capacity to cut other stones. As gems in their own right, however, they had to be appreciated in their natural octahedral state as methods to cut them would not be developed until the late XIV century. At that time, the

Gems

Gemstones must be cut properly to maximize sparkle and brilliance

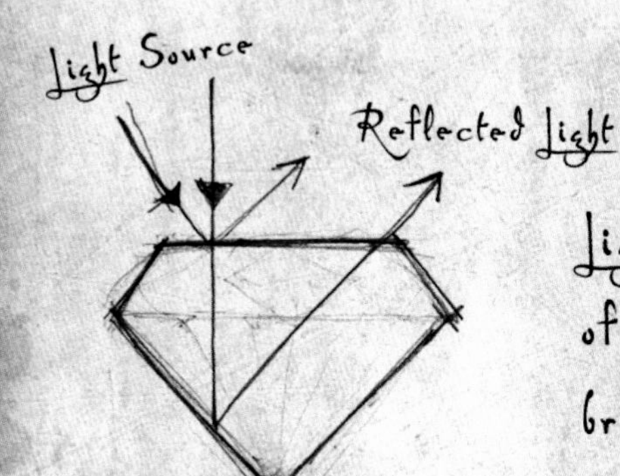

Light bounces off the facets of the stone resulting in brilliance and sparkle

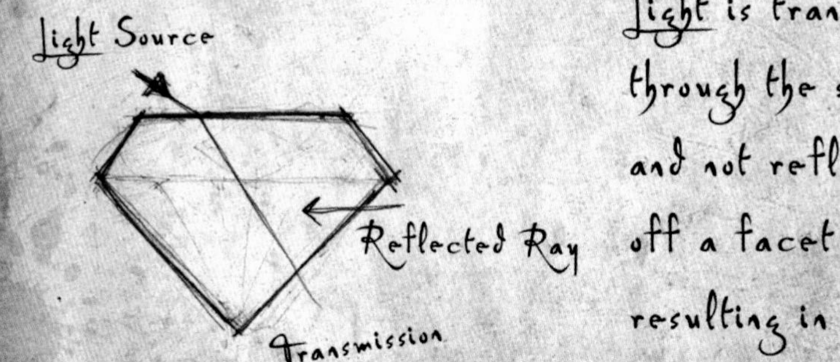

Light is transmitted through the stone and not reflected off a facet resulting in a light, but flat colored stone

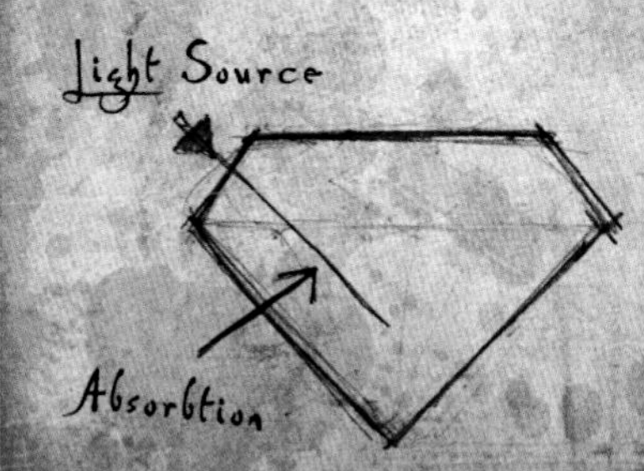

Internally reflected and thus absorbed resulting in a dull stone

BLACK DIAMONDS

Caterina de' Medici (1519-1589) was actually diamond encrusted, but you wouldn't know it. Diamonds, the hardest substance on Earth could not be cut easily during her era. They were set in their natural octahedral shape or cut with rudimentary methods which yielded poor brilliance. Artists thus portrayed them as black. Note the black stones in her pectoral brooch and junctures between strands of pearls, they are diamonds.

art of diamond cutting began to flourish, initially in Venice and then in the Low Countries. Major advances were made throughout the Renaissance, as gemology began a slow transformation from a purely mercantile practice of procuring rare and costly items to the technology-based science firmly rooted in the principles of geology, geochemistry and physics we know today.

Literary Gems

The Late Middle Ages saw a florescence of gemological literature, much of which elaborated upon the well-known works of Classical antiquity such as Pliny's late I century A.D. *Naturalis historia* (Natural History) in which he devotes Book XXXVII to gemology. Despite the introduction of numerous errors caused by centuries of hand-copying, this work remained the standard text on the subject for well over a thousand years. It is quite possible that antiquaries in Dante's day had access to Theophrastus' IV century B.C. *De lapidibus* (On Stones), a source upon which Pliny relied and chapters of which may have been in circulation at the time, albeit likely in the original Greek.

Dante would have been familiar with such lapidary treatises as *Liber lapidum seu de gemmis* (ca. 1061-1081), a poem in written in Latin hexameters by Marbodius of Rennes (ca. 1035–1123)—better known as Marbode—in which he describes some sixty stones and their magical properties. Neither the Bible nor the Fathers of the Church are ever quoted in the work; the only authors mentioned in the text are pagan. It may seem strange that so secular a poem should be the work of a bishop, but its subject matter is derived from Classical authors and Marbode may have written it as an exercise in elegant verse rather than as a serious treatise. Regardless, the work became renowned and was soon translated into Italian, French and many other languages and became the model upon which subsequent poetic works with lapidary themes were written by Dino Compagni, Bencivenni Zucchero and Francesco Sacchetti.

Many anonymous Italian lapidaries wrote in Latin, but one work, *De mineralibus*, was an important scientific treatise by the Dominican friar, Albertus Magnus (ca. 1206–1280), who is among the dozen intellectuals Dante "meets" in his fourth sphere of Paradise. The poet likely had knowledge of other manuscripts, including the encyclopedic *Speculum naturale* (Mirror of Nature), by the Dominican friar Vincent of Beauvais (ca. 1190-1264) and Bartholomaeus Anglicus' *De proprietatibus rerum*, a work detailing the medicinal properties of stones penned sometime in the early XIII century. According to geologist Annibale Mottana of the Università degli Studi Roma Tre, another influential work of the time was the *Lapidarius*, a volume of dubious origin, which, he contends, contributed greatly to the transfer of information on precious stones from Arab countries to

Europe. Long attributed to Aristotle and translated into Latin by Alfredus Anglicus ca. 1200 A.D., the *Lapidarius*, says Mottana, was most likely written by an VIII century Syrian monk, whose work was later elaborated upon and embelished by Arab, Hebrew and Latin translators.

While varying in their scientific content, most of these works include details regarding the medicinal properties of various gems and their astronomical associations. Certain jewels were considered lucky or unlucky for an individual depending the sign of the zodiac under which he was born and its attendant birthstone. It was common at the time to cast a person's horoscope through the observation of the celestial conditions that prevailed at the time of birth in order to determine their future. Despite multiple ecclesiastical condemnations of the practice, it continued to gain popularity among intellectuals during the Late Middle Ages and Early Renaissance with the increasing availability of recently translated Arabic astrological texts entering Europe. Among these was the influential *Liber Astronomiae* (ca. 1277) by the celebrated astrologer Guido Bonatti, whom Dante places among heretical soothsayers in Canto XX of his *Inferno*.

> *Vedi Guido Bonatti; vedi Asdente,*
> *ch'avere inteso al cuoio e a lo spago*
> *ora vorrebbe, ma tardi si pente.*
>
> **Behold Guido Bonatti, behold Asdente,**
> **Who now unto his leather and his thread**
> **Would fain have stuck, but he too late repents.**
> *(Inferno, XX, 118-120)*

Other stones were worn as special talismans as they had a rich history of religious and numerologic symbolism. Especially prevalent was the reference to "twelves"—apostles, Tribes of Israel,

Foundation Stones of the New Testament, jewels on the high priest's breastplate in the story of Exodus and the angels thought to guard the Gates of Paradise. Within each of these groups, a special stone was assigned to each of the twelve and promised the wearer of a specifically chosen gem, abundance and good fortune.

In addition to these Biblical associations, Dante was no doubt familiar with the abundant lore that surrounded many precious stones—fantastic tales of gemmed cities of the gods such as that described by the II century A.D. satirist Lucian in the second book of his *Vera historia* (A True Story). There were also rumors of "diamond rivers" in India, no doubt a reference to the fact that all diamonds in Dante's time were from alluvial deposits. Mining of diamonds found in kimberlite and lamproite only began in the 1870s, following the discovery of South Africa's vast diamond fields.

Many of Dante's metaphors are based on the unique reflective and refractive properties intrinsic to all gems. We now know that his poetic "illuminated" imagery was based on his understanding of the physics of light. He was most likely influenced by the ancient texts which were starting to emerge from monasteries and universities in the 1200s. Among them, particularly diffuse in Venice, were treatises on Euclidean and Pythagorean geometry which Italian gem cutters used to resolve complex problems such as the determination of angles, symmetry and the effect of light on a facet.

In his use of gems as critical elements the *Divine Comedy*, Dante takes into account all of the astronomical, theological and mythological knowledge that was available in his day. Particularly in the *Paradiso*, using the interplay of light on jewels, he creates literary images worthy of Paradise. For within this heavenly realm, cut gemstones symbolize the revelation of the soul after the dross of the body has been chipped away, only then can the soul's sparkling facets reflect divine light.

⋆ The Gems of Paradise ⋆

PEARLS

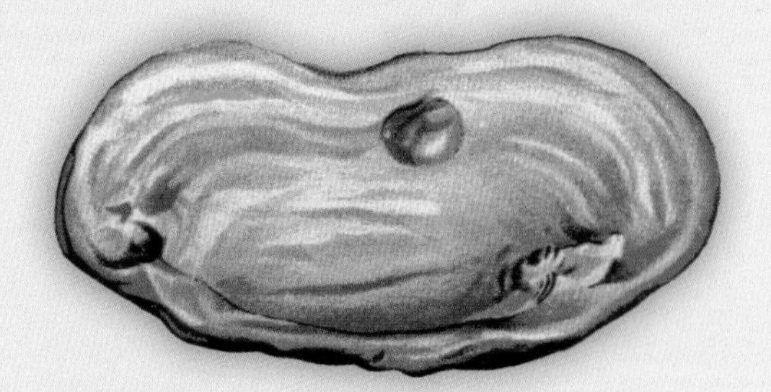

No other object used for personal adornment has held its comparative position of desirability for as long as the pearl. According to Pliny, *Naturalis historia* 9.54, "the very highest position among all valuables belongs to the pearl." The origin of the name pearl can be traced through a number of languages: in Latin *perna*, and in Greek *margarite*, which may have derived from the Babylonian for "child of the sea." One reason that pearls have been so widely used since antiquity is that they require no fashioning to reveal their beauty. They are mentioned in the *Book of Job* and the *Talmud*. The Egyptians, Persians and Hindus held them in great esteem. It was the favorite gem among the rich of the Roman Empire where both men and women used them lavishly, not only for personal adornment but for embellishing their domestic surroundings. It is reported that Roman women wore them in their sleep to remind them of their wealth when they awoke.

Pearls brought back from the Orient by the Crusaders probably increased the appreciation of these gems in Europe. John of St. Amand wrote in the second half of the XIII century that pearls comfort the heart by similarity, since they are hard—like the heart.

Until the XVII century, pearls were used in Europe for a host of medicinal purposes—from aphrodisiacs to cures for insanity

Pearl

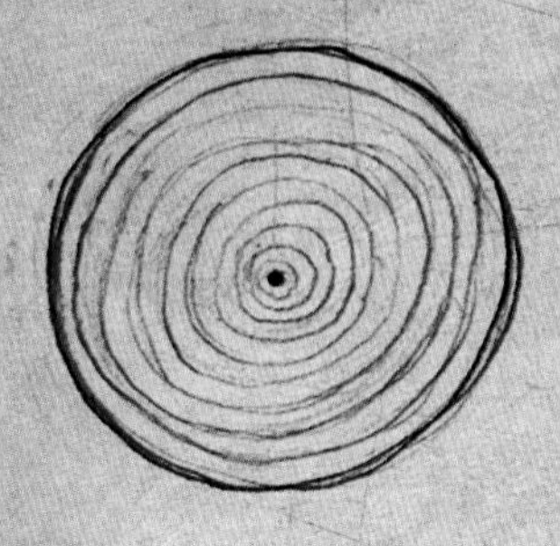

Layers of calcium carbonate (nacre) grow slowly around an internal seed or irritant. The multiple layers build up, resulting in a luminous depth of color called lustre.

The color of pearls is the result of two separate components: body color (for background) and overtone. The more layers of aragonite platelets, called nacre, the more lustrous the pearl. The color of the pearl is due to the type of oyster in which it grows and trace elements in the acquatic environment.

and indigestion—a practice that had its origins in the East. They were also thought to detoxify the system and restore youth. When pulverized, they were made into potions and balms formulated to treat many ailments.

The actual pearl formation process was long steeped in mystery. Indians believed that pearls were formed when heavenly drops of dew, which fell into the sea, were caught by an open mollusk under the first rays of the sun during a full moon. In good weather, a lovely pearl was formed, in poor weather a dark, less attractive one resulted. According to Hebrew scripture, pearls are the remains of tears shed by Eve when she was banished from the Garden of Eden.

To form a round pearl, the mollusk deposits layers of overlapping crystals or platelets of a carbonate mineral, aragonite ($CaCO_3$)— called "nacre"—concentrically around a nucleus. This positioning of the crystals and platelets gives rise to the diffraction and interference of light that produces iridescence—called "orient" when seen on the surface of a pearl. Differences in the number and thickness of these layers are responsible for the different qualities of orient and thus, translucency of pearls. The greater the number of nacreous layers below the surface, the more lustrous the pearl.

The color of most pearls is the result of two separate components: body color (for the background) and overtone. When present, the overtone will be superimposed on the body color. The body color is most easily distinguished visually from the overtone under soft, diffused light. The overtone is seen in the actual reflection of the light source from the surface of the pearl and the remainder of the pearl shows the body color.

In several passages of the *Divine Comedy*, Dante describes the pearl much as a modern pearl dealer might: lustrous, large, perfect, creamy, translucent and iridescent.

> ***Per entro sé l'etterna margarita***
> ***ne ricevette, com' acqua recepe***
> ***raggio di luce permanendo unita.***
>
> Into itself did the eternal pearl
> Receive us, even as water doth receive
> A ray of light, remaining still unbroken.
>
> *(Paradiso, II, 34–36)*

Due to its roundness and crystalline structure, a pearl can receive light from many directions and return unbroken reflections as a function of its light dispersion properties, which are unlike any other gem's. A faceted gem is much more sensitive to the light direction for reflection.

> ***e la maggiore e la più luculenta***
> ***di quelle margherite innanzi fessi,***
> ***per far di sé la mia voglia contenta***
>
> And now the largest and most luculent
> Among those pearls came forward, that it might
> Make my desire concerning it content.
>
> *(Paradiso, XXII, 28–30)*

In this passage, the largest and most lustrous of the pearls, the precious spirit of St. Benedict, comes forward to talk to Dante. As the father of monasticism in the Western Church, St. Benedict exemplifies the epitome of contemplative holiness—which stood in sharp contrast to the corruption of the Church Dante witnessed in his day. Because the pearl's surface is round, it reflects light in a manner in which it appears to emanate from within. The soft radiance of a large, valuable, lustrous pearl is a fitting choice for such a mighty Christian leader.

Per la natura lieta onde deriva,
la virtù mista per lo corpo luce
come letizia per pupilla viva.
Da essa vien ciò che da luce a luce
par differente, non da denso e raro;
essa è formal principio che produce,
conforme a sua bontà, lo turbo e 'l chiaro.

From the glad nature whence it is derived,
The mingled virtue through the body shines,
Even as gladness through the living pupil.
From this proceeds whate'er from light to light
Appeareth different, not from dense and rare:
This is the formal principle that produces,
According to its goodness, dark and bright.
(Paradiso, II, 142–148)

Dante, having entered the sphere of the Moon, which he likened to an eternal pearl, did not understand how spots could have appeared on the diamond-smooth surface of this celestial body. Beatrice explains that they are optical phenomena caused by divine light. In reality, both the Moon and a pearl are not perfect. We can see the Moon's surface relief, but it is only when one passes a string of pearls gently over one's teeth that miniscule imperfections can be noticed. It is these natural marks which prove a pearl is real.

Quali per vetri trasparenti e tersi,
o ver per acque nitide e tranquille,
non sì profonde che i fondi sien persi,
tornan d'i nostri visi le postille
debili sì, che perla in bianca fronte
non vien men forte a le nostre pupille;

Such as through polished and transparent glass,
Or waters crystalline and undisturbed,
But not so deep as that their bed be lost,
Come back again the outlines of our faces
So feeble, that a pearl on forehead white
Comes not less speedily unto our eyes;
(Paradiso, III, 10–15)

Modern merchants make numerous distinctions of light and color to determine the value of a pearl. How light plays off its surface gives it a higher value. Evaluators must be able to make the subtlest of distinctions of light vs. light and light vs. shadow to determine the quality of the pearl. John Freccero, in his *Introduction to the Paradiso*, speaks to the fact that, "So with the poem, which manages to approach its conclusion and silence by the gradual dissipation of all difference between light and light, and yet remains as the shadow of all that the experience is not, as irreducibly literary as 'a pearl on a white brow'… the interplay of light and shadow as a figure for the poem itself."

Dante shows absolute genius in selecting the pearl as an emblem for the poem. Pearl-like objects are found in many mollusks, but very few have the orient and iridescence of a true pearl. Those without a build-up of nacre are known as calcareous concretions and are lusterless and without value as gems. And so, Dante creates a fine distinction in the *Paradiso*—it takes light to distinguish the priceless from the worthless, a subtle bit of iridescence makes all the difference in the world.

⋆ The Gems of Paradise ⋆

TOPAZ

According to Pliny (*Naturalis historia*, 37.32) the topaz was named for Topazos, an island in the Red Sea that was often obscured by fog and therefore difficult to find. The name he says, comes from the word 'topazin' meaning 'to seek,' in the language of the Troglodytæ. Others contend that the name is a derivative of the Sanskrit word *topas* meaning fire.

During the Middle Ages in Europe, topaz was not particularly popular, although it was occasionally used in ecclesiastical or royal jewelry. Perhaps no other gem has had more varied or preposterous powers ascribed to it. In fact, when topaz was worn as an amulet, it was said to drive away sadness, dispel enchantment, strengthen the intellect, prevent mental disorders, bestow courage and prevent sudden death. When the stone was pulverized and put into wine, it was used as a cure for asthma, insomnia, burns and hemorrhage. Marbode's poetic treatise recommends it as a cure for weak vision. His prescription called for immersing the gem in wine for three days and nights, then placing it on the afflicted eye. According to Ragiel's XIII century *The Book of Wings*, a topaz engraved with the figure of a falcon could help its bearer cultivate the goodwill of kings, princes and magnates. Topaz was recommended by Geronimo (*Gerolomo*) Cardano as a cure for madness, a means of increasing one's wisdom and prudence and a coolant for both boiling water and excessive anger. All of these powers were believed to increase and decrease with changes of the Moon. Because of its low light refraction index of 1.6, topaz can often be mistaken for glass. This same refraction property makes it excellent for optical purposes.

Ben supplico io a te, vivo topazio
che questa gioia preziosa ingemmi,
perché mi facci del tuo nome sazio.

Truly do I entreat thee, living topaz!

Set in this precious jewel as a gem,

That thou wilt satisfy me with thy name.

(Paradiso, XV, 85–87)

In this canto, the souls are mounted like gems in a cross of light. Here Dante refers to his revered great-great-grandfather, the crusader Cacciaguida degli Elisei (ca. 1091–1148) as the "living topaz," thus distinguishing him as a larger gem in this shimmering cross. This term of endearment is the medieval equivalent of "a chip off the old block."

Anche soggiunse: "Il fiume e li topazi
ch'entrano ed escono e o'l rider de l'erbe
son di lor vero umbriferi prefazi.
Non che da sé sian queste cose acerbe;
ma è difetto da la parte tua,
che non hai viste ancor tanto superbe."

She added: "The river and the topazes

Going in and out, and the laughing of the herbage,

Are of their truth foreshadowing prefaces;

Not that these things are difficult in themselves,

But the deficiency is on thy side,

For yet thou hast not vision so exalted."

(Paradiso, XXX, 76–81)

In this pastoral setting, Dante uses gems, minerals and colors to eliminate the concept of mortality by changing living objects, which will die, into enduring ones. The river, the symbol of living grace is transformed into a stream of light with sparks emanating from the gems. The topazes are symbols of the angels which are ministering to the souls (flowers) which have been transformed into rubies set in gold. These gems portray in vivid color the beauty of springtime without fading or dying.

⋆ The Gems of Paradise ⋆

SAPPHIRE

From ancient times, the sapphire, which comes from *sappheiros,* or "blue stone" in Greek, was associated with curing illnesses of the eyes. Although it must be noted that Classical authors such as Pliny used the term sapphire to describe lapis lazuli rather than the precious gem that now bears this name. The sapphire was often used to cure plague boils. A gem of a fine, deep color was selected and rubbed gently and slowly around the sore. A good while after the removal of the stone, the patient's boil would disappear.

Marbode lavishes praise upon this beautiful stone, for it was in his time that it was chosen as the gem most appropriate for ecclesiastical rings. Marbode further noted that the sapphire's powers included capabilities of banishing fraud and preventing terror and poverty. Necromancers honored the sapphire as it enabled them to hear and understand the most obscure oracles.

The stone also has heavenly connotations. A medieval French text (circa 1265) on the significance of the sapphire states: "...for the contemplation of sapphires should raise men's souls to the contemplation of the heavenly kingdom..."

nous senefie
L'esperance ou preudom se fie
Qui garde au haut regne celestre,

The star sapphire, which exhibits a six-rayed star caused by needle-like inclusions (often of the mineral rutile) within the gem, has been known since antiquity as the "Stone of Destiny," its three crossed lines are thought to represent faith, hope and destiny. Still another legend refers to these gems as sparks from the Star of Bethlehem, the reason being, as the stone is moved, or the light changes, a star appears, shines and moves due to a phenomena called asterism. Here Dante shows the light following along the axes of the star of the gem creating a glowing cross.

né si partì la gemma dal suo nastro,
ma per la lista radial trascorse,
che parve foco dietro ad alabasto.

Nor was the gem dissevered from its ribbon,

But down the radiant fillet ran along,

So that fire seemed it behind alabaster.

(Paradiso XV, 22–24)

In fact this star-like image was thought to be the key to warding off ill omen and the evil eye. As a talismanic stone, it was said to be so potent that it continued to exercise its positive influence over the first wearer even when it passed into other hands.

The poet's creativity continues as he invents new words. This process, called neologism, is evident in *"ingemmarsi"* (to bejewel) and "*inzaffirarsi*" (to adorn with sapphires). These words not only evoke glorious images, but contributed to constructing an entirely new language that was refined and innovative.

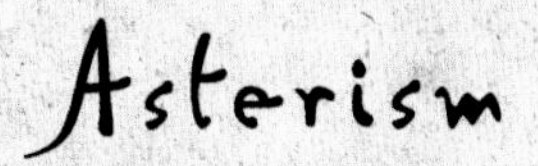

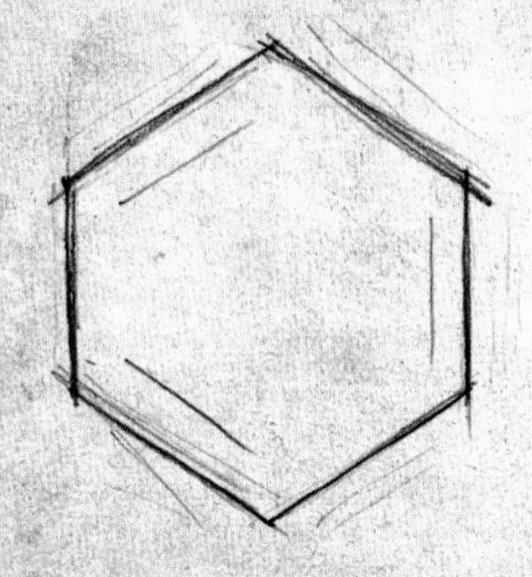

Corundum has six faces.
Rutile needles form
parallel to the crystal axes

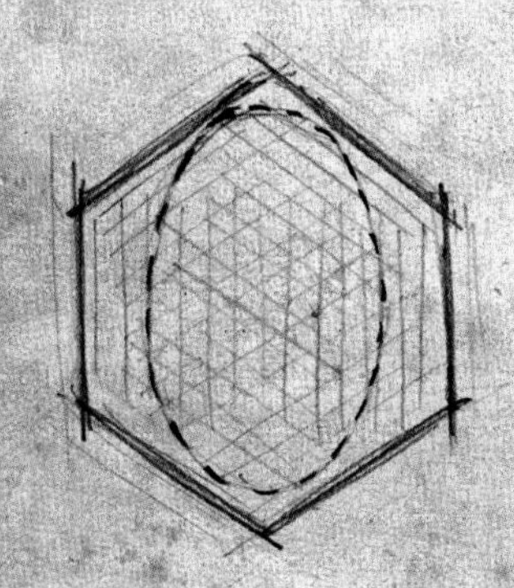

The rutile needles grow parallel
to the crystal axes and result
in asterism.
The outline shows how a cabochon
should be oriented and cut

A star results when adirect light
source reflects off the oriented
needles

Qualunque melodia più dolce suona
qua giù e più a sé l'anima tira,
parrebbe nube che squarciata tona,
comparata al sonar di quella lira
onde si coronava il bel zaffiro
del quale il ciel più chiaro s'inzaffira.

Whatever melody most sweetly soundeth
On earth, and to itself most draws the soul,
Would seem a cloud that, rent asunder, thunders,
Compared unto the sounding of that lyre
Wherewith was crowned the sapphire beautiful,
Which gives the clearest heaven its sapphire hue.
(Paradiso, XXIII. *100–102)*

In this canto, Dante creates a neologism as he turns the noun *zaffiro* (sapphire) into the verb *s'inzaffira* (to adorn with sapphires), referring to the sky's blue color. The sapphire is mentioned earlier in the *Divine Comedy*, in the *Purgatorio*, where he notes "the sweet hue of the oriental sapphire," which he saw gather in the heavens. A reference not only to its geographical origin in the East, but perhaps also to emphasize that as men approach the blue sky, which he uses as the symbol of the Blessed Virgin, grace comes to them. The use of the blue sapphire as a symbol of the Virgin Mary also seems to be Dante's own invention and has endured to this day.

He continues with his use of gems as verbs in the subsequent sections of the *Paradiso*, as if he needed to create the exact words to illustrate the brilliance of a divine world so different from our own. His creativity and facility in contributing to a new language through the invention of words and grammatical acrobatics is unparalleled. In this way he influenced generations of future poets and assured immortality for his own works.

O dolce stella, quali e quante gemme
mi dimostraro che nostra giustizia
effetto sia del ciel che tu ingemme!

O gentle star! what and how many gems
Did demonstrate to me, that all our justice
Effect is of that heaven which thou ingemmest!

(*Paradiso*, XVIII, 115–117)

Poscia che i cari e lucidi lapilli
ond' io vidi ingemmato il sesto lume
puoser silenzio a li angelici squilli,

After the precious and pellucid crystals,
With which begemmed the sixth light I beheld,
Silence imposed on the angelic bells,

(*Paradiso*, XX, 16–18)

STAR SAPPHIRE

The Star of India, at 563.35 carats, is probably the largest gem of its kind in the world. The star is formed by needle-like crystals, normally of rutile, which criscross in the stone and reflect the light in a star pattern. The gem is on display, in all its glory, at the Museum of Natural History in New York.

* The Gems of Paradise *

DIAMOND

The word diamond comes from the ancient Greek *adámas*, meaning unalterable, unbreakable or untamed. For millennia, in India, diamonds were found along the banks of the Penner, Krishna and Godavari Rivers, collected by searching carefully in the alluvial deposits along the banks. Diamonds were revered and highly valued as talismans as far back as 800 B.C. and for more than 2,500 years India was the only supplier of the gemstone to the world. Diamonds and their reputation for metaphysical powers arrived in Rome around the I century B.C. Pliny (*Naturalis historis* 37.15) tells us that *"adamas… overcomes and neutralizes poisons, dispels delirium, and banishes groundless perturbations of the mind."* He describes them as stones that will scratch all others, confirming the name for the hardest substance, adamas. He adds that *"the substance that possesses the greatest value, not only among the precious stones, but of all human possessions, is adamas; a mineral which, for a long time, was known to kings only, and to very few of them."*

Interestingly, diamonds, while known to the Romans, were actually quite rare in Europe until the XIV century. Indeed, very few kings possessed them. Louis IX of France (1214–1270) issued an order forbidding women, including queens to wear them. Even then, most diamonds were prized for their rarity and capacity to cut other stones rather than as gems in their own right. The virtues ascribed to the diamond are almost all directly traceable either to its unconquerable hardness or to its transparency and purity. It was thought to bring victory to the wearer by endowing him with superior strength, fortitude, and courage. St. Hildegard states that the sovereign virtue of the diamond was recognized by the devil who was a great enemy of this stone because it resisted his power by day and by night.

Diamonds depicted in Medieval and Renaissance art often appear as black stones due to their uncut state, the rudimentary lapidary techniques which were just developing or possibly a visual commentary on the inferior stones imported at the time. It would be nearly a half century after Dante's time however, that the superior brilliance of the diamond would start to be more widely diffuse through faceting, a technique discovered in Venice, which would transform them into the sparkling gems we enjoy today.

Of course, this is not to say that brilliant uncut diamonds were unknown, they were just extremely rare. Dante would have had the pleasure of seeing them as he describes his entrance into the sphere of the moon and encounters Beatrice. He likens the experience to entering into a glistening diamond, a stone whose durability and perfection were unparalleled.

Parev' a me che nube ne coprisse
lucida, spessa, solida e pulita,
quasi adamante che lo sol ferisse.

It seemed to me a cloud encompassed us,
Luminous, dense, consolidate and bright
As adamant on which the sun is striking.
(Paradiso, II, 31–33)

VERMEER'S CATHOLIC CONVERSION

One of his few religious paintings, *Allegory of the Catholic Faith* was painted after Vermeer married a Catholic woman and adopted her faith. The transparent crystal sphere references the perfection and immaculate nature of the divine, while the pearls in the woman's necklace are ancient symbols of purity.

This image of a brilliant, stainless sphere would capture the imagination of future artists. In 1670, Vermeer, who had converted to Catholicism, would use gems as symbols in his paintings. In *Allegory of the Catholic Faith* (1670–72) he depicted the Heavenly realm as a clear adamantine sphere.

> ***Di questa luculenta e cara gioia***
> *del nostro cielo che più m'è propinqua,*
> *grande fama rimase; e pria che moia,*
> Of this so luculent and precious jewel,
> Which of our heaven is nearest unto me,
> Great fame remained; and ere it die away
> (*Paradiso*, IX, 37–39)

In this canto, Cunizza—the sister of Ezzelino, a Ghibelline tyrant who appears in Dante's seventh ring of Hell—praises Folco, a famous troubadour who became a monk and later a bishop, as a brilliant jewel.

⋆ The Gems of Paradise ⋆

RUBY

The powers that have been ascribed to ruby over the centuries are innumerable, for it controlled bodily and mental health, evil thoughts, amorous desires and aided in settling disputes. The early Burmese thought the stone would bestow invulnerability when it was actually inserted into the owner's flesh.

The glowing hue of the ruby suggested the idea that an inextinguishable flame burned in this stone. In the *Lapidaire en Vers* of Philippe de Valois (1293-1350), it is said that *"the books tell us the beautiful clear and fine ruby is the lord of stones; it is the gem of gems, and surpasses all other precious stones in virtue."* He continues by calling it *"the most precious of the twelve stones God created when he created all creatures."* By Christ's command, the ruby was placed on Aaron's neck *"the ruby, called the lord of gems, the highly prized, the dearly loved ruby, so fair with its gay color."*

In the *Divine Comedy*, Dante uses rubies to describe Christian warriors and souls of the blessed.

L'altra letizia, che m'era già nota
per cara cosa, mi si fece in vista
qual fin balasso in che lo sol percuota.

The other joy, already known to me,
Became a thing transplendent in my sight,
As a fine ruby smitten by the sun.
(*Paradiso*, IX, 67–69)

This passage is particularly important as Dante chooses not to use the Italian word for ruby, *rubino* or *rubinetto* as he does later passages, but rather *balasso*, displaying his encyclopedic knowledge of gems and his compulsion to expand the Italian language by creating a new word, a poetic Italianization of the Arab word *balaksh*, which referred to a rare and special type of ruby prized for its violet-rose hue. While called a ruby, it may have been a spinel which came from the ancient ruby and spinel mines of Badakhshan, Afghanistan, from which the shortened name, "balas" is derived. The mines were famous since ancient times as they were also the source of the precious gem, lapis lazuli.

Returning from his trip to the East, Marco Polo (1254-1342) commented on the mines which in the future would provide monarchs with enormous red, sparkling gems called Balas Rubies. It would only be in 1783 that a French mineralogist would classify spinel as a mineral separate from ruby, and thus render many crown jewels much less valuable, but nonetheless still stunningly beautiful.

parea ciascuna rubinetto in cui
raggio di sole ardesse sì acceso,
che ne' miei occhi rifrangesse lui.

Appeared a little ruby each, wherein
Ray of the sun was burning so enkindled
That each into mine eyes refracted it.
(*Paradiso*, XIX. 4–6)

Here, Dante describes the image of divine justice with the congregated souls, each like an individual little ruby, but reflecting singly the sun's rays. All the souls in this sphere speak collectively not individually. Dante describes them as flowers, an image continued in the following passage.

> *Di tal fiumana uscian faville vive,*
> *e d'ogne parte si mettien ne' fiori,*
> *quasi rubin che oro circunscrive;*
>
> Out of this river issued living sparks,
> And on all sides sank down into the flowers,
> Like unto rubies that are set in gold;
> *(Paradiso, XXX, 64–66)*

The enduring beauty of the souls, which will continue to sparkle and retain their color and inner fire is exemplified in the use of gems as a vehicle for transforming the transient, fading beauty of life into the everlasting beauty of the soul as it is transformed by divine light.

⁎ The Gems of Paradise ⁎

EMERALD

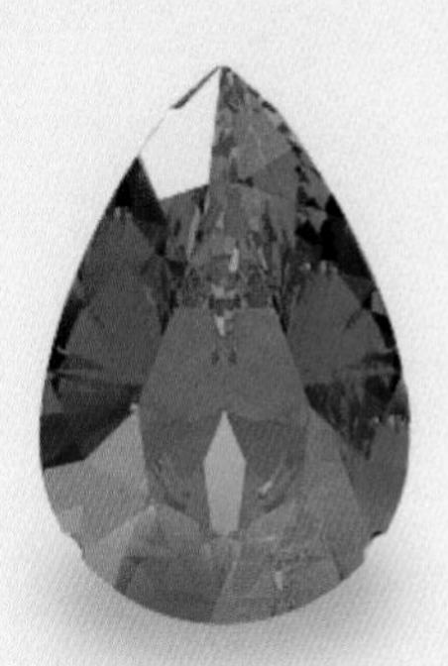

Dante invokes the image of an emerald only once in the *Divine Comedy*, in the *Purgatorio*. In this canto, Dante praises its color, especially when it is moistened, noting that the greens, blues and silvers of gems and minerals are everlasting.

> *Oro e argento fine, cocco e biacca,*
> *indaco, legno lucido e sereno,*
> *fresco, smeraldo in l'ora che si fiacca,*
>
> **Gold and fine silver, and scarlet and pearl-white,**
>
> **The Indian wood resplendent and serene,**
>
> **Fresh emerald the moment it is broken,**
>
> *(Purgatorio, VII, 73-75)*

The word emerald comes from the Greek *smaragdos*, which means "green gem." Its name, however, was likely derived from earlier Hebrew or Sanskrit. Emeralds found at Pompeii and Herculaneum likely came from mines at Zabara, on the Red Sea coast in Upper Egypt, which had been worked since the time of Cleopatra (69-30 B.C.).

Pliny (*Naturalis historius* 37.16) describes the emerald in the following manner: *"And then, besides, of all the precious stones, this is the only one that feeds the sight."* Pliny's observations had their antecedents in Classical Greek texts such as that of Theophrastus in the IV century B.C. where he noted emerald's power to soothe the eyes.

For the Romans, the emerald symbolized the reproductive forces of nature and was thus associated with Venus. Yet in ancient times it was also considered an adversary of sexual passion. In the XIII century, Albertus Magnus wrote that when King Bela of Hungary (1235-1270) embraced his wife, his magnificent emerald broke into three pieces.

In the XI century, Marbode noted that an emerald improves memory, makes its owner eloquent and persuasive and brings him great joy. As a revealer of truth, this stone was an enemy of all enchantments hence, it was greatly feared by magicians who found all their arts of no avail if an emerald were in the vicinity when they began to weave their spells.

The Gems of Paradise

GLASS, AMBER & CRYSTAL

With regard to glass, amber, and crystal—which are not gems—Dante clearly understood the physics of light in that these three substances are amorphous, i.e. lacking in internal crystal structure, and therefore without refraction; light simply passes through them.

> *E come in vetro, in ambra o in cristallo*
> *raggio resplende sì, che dal venire*
> *a l'esser tutto non è intervallo,*
>
> And as in glass, in amber, or in crystal
> A sunbeam flashes so, that from its coming
> To its full being is no interval,
>
> *(Paradiso,* XXIX, *85–27)*

> *Poscia tra esse un lume si schiarì*
> *sì che, se 'l Cancro avesse un tal cristallo,*
> *l'inverno avrebbe un mese d'un sol dì.*
>
> Thereafterward a light among them brightened,
> So that, if Cancer one such crystal had,
> Winter would have a month of one sole day.
>
> *(Paradiso,* XXV, *100–102)*

From December 21–January 21, the constellation, Cancer, the crab, rises at sunset and sets at sunrise. One soul, St. John, presented as a luminous crystal, stood out so much that the glow of his star, if inserted into the constellation, would be so powerful that it would illuminate the sky for one month.

Dante's use of gems in the *Paradiso* shows his knowledge of the subject was thorough. What is so moving, however, is his consummate display of poetic genius in turning gems, beautiful in and of themselves, into literary gems—images and metaphors so powerful that they have been handed down through the ages to provide a feast for the mind and spirit.

APPENDIX *Order in the Universe*

Despite Dante's terracentric view, it is clear that the order in which these bodies existed in relation to the Earth was understood at some level.

First	The Moon	Realm of the Inconstant
Second	Mercury	Realm of the Ambitious
Third	Venus	Realm of Lovers
Fourth	The Sun	Realm of the Wise
Fifth	Mars	Realm of the Warriors of the Faith
Sixth	Jupiter	Realm of the Just Rulers
Seventh	Saturn	Realm of the Contemplatives
Eighth	The Fixed Stars	Realm of Faith, Hope and Love
Ninth	The Primum Mobile	Realm of Angels
Tenth	The Empyrean	Abode of God

APPENDIX *Order in the Universe*

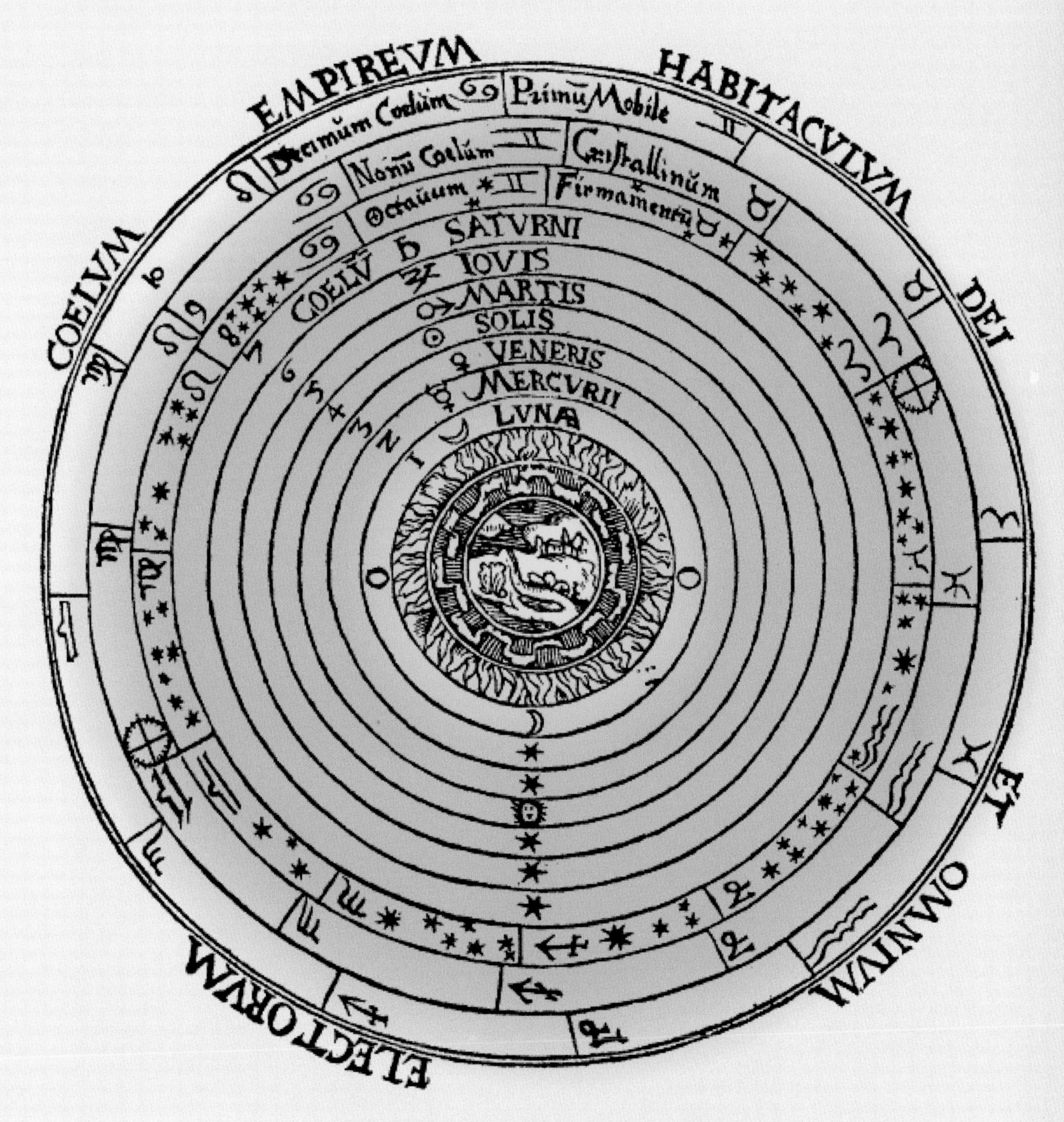

DANTE'S EARTH CENTERED UNIVERSE

The planets and fixed stars are found in rotating spheres nested one inside of another. The heavenly bodies are embedded in an ethereal, transparent atmosphere, much like jewels set in orbs. The *Divine Comedy* was instrumental in presenting this model to the medieval audience.

APPENDIX *Notions of Angelic Order*

Passages within Canto XXVIII of the *Paradiso* are focused in part on a discussion of angelic orders. Christian tradition organized angels in ranks, with their various functions outlined in Paul's Epistles. In this canto, Dante questions the great Pope, St. Gregory, about the proper order. For Dante, references to the angels were representative of Divine Providence and the spiritual order that governs our lives.

List of stones pertaining to the nine orders of angels according to Gregory the Great:

GEM	ORDER OF ANGELS
Sard	Seraphim
Topaz	Cherubim
Jasper	Thrones
Chrysolite	Dominions
Onyx	Principalities
Beryl	Powers
Sapphire	Virtues
Carbuncle	Archangels
Emerald	Angels

APPENDIX *Apostolic Stones*

Foundation gemstones mentioned in *The Book of Revelation*, XXI, 14, were inscribed with the names of apostles. In medieval times, this book became the subject of intense study for mystical references. Scholars sought to associate each apostle with a stone. Varying points of view however, resulted in the fact that the lists don't always match. Some theologians were opposed to the idea of assigning foundation stones to the apostles for they held that only Christ himself could be regarded as the foundation of His Church.

MONTH	APOSTLE	GEM	VIRTUE
January	St. Peter	Jasper	Satisfaction
February	St. Andrew	Carbuncle	Passion of Christ
March	St. James & John	Emerald	Kind, good
April	St. Philip	Carnelian	Shed blood
May	St. Bartholomew	Chrysolite	Excellence
June	St. Thomas	Beryl	Moderation
July	St. Matthew	Topaz	Uprightness
August	St. James the lesser	Sardonyx	Strength
September	St. Thaddeus	Chrysoprase	Sternness to sin
October	St. Simeon	Jacinth	Royal dignity
November	St. Matthius	Amethyst	Perfection
December	St. Paul	Sapphire	Soul

APPENDIX *Of Tribes and Astral Signs*

In the Middle Ages, Hebrews associated each of the 12 Tribes of Israel with a zodiacal sign and gemstone. This would later develop into the tradition of birthstones.

SIGN	GEM	TRIBE
Aquarius	Garnet	Naphtali
Pisces	Amethyst	Asher
Aries	Bloodstone	Judah
Taurus	Sapphire	Issachar
Gemini	Agate	Zebulun
Cancer	Emerald	Reuben

SIGN	GEM	TRIBE
Leo	Onyx	Simeon
Virgo	Carnelian	Gad
Libra	Chrysolite	Ephraim
Scorpio	Beryl	Manasseh
Sagittarius	Topaz	Benjamin
Capricorn	Ruby	Dan

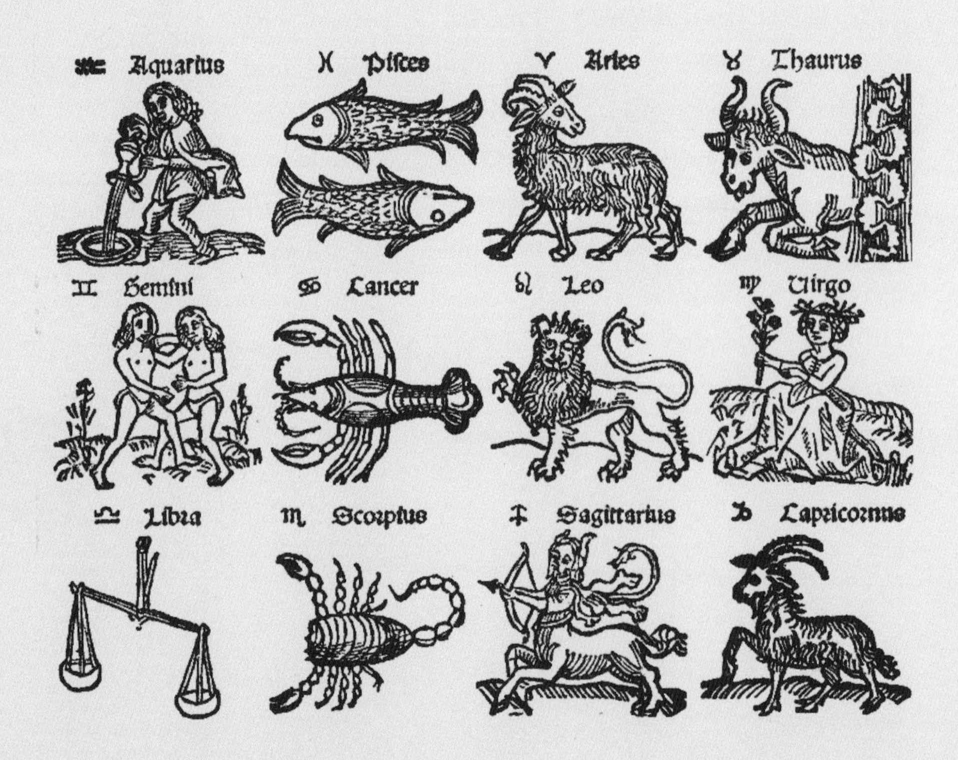

APPENDIX *Of Tribes and Astral Signs*

MARBODAEI
GALLI CAENOMANENSIS DE
gemmarum lapidumq́; pretioſorum formis,natu
ris,atq́uiribus eruditū cū primis opuſculū,ſane q̄utile,cum
ad rei medicæ,tū ſcripturæ ſacræ cognitionē:nūc primū nō
mō cētū ferme uerſib.locupletatū pariter& accuratius emē-
datū, ſed & ſcholijs q̄q; illuſtratū p Alardū AEmſtelredamū

Cuius ſtudio additę ſunt & præcipuæ gemmarū lapidūq; ptioſorū explicatiões, ex uetuſtiſſ. qbuſq; autorib' coactæ. Cū ſcholijs Pictorij Villingeñ.

Ἐν μαργάριτον τίμιον. Ἀποδοὺς ἅπαντα λάμβανε.
En margaritū nobile, Eme ſi cupis diteſcere.
Rationale. Exodi. 28.& 39. Leui.8.

Doctrina & Veritas

3 Smaragdus Leui	4 Carbūcul'. Iuda	5 Saphirus Zabulon
2 Topazius Simeon	9 amethyſtus Aſer	6 Iaſpis Iſachar
1 Sardius Ruben	10 chryſolitus Nepthalim	7 Lincurius Dan
12 Beryllus Beniamin	11 Onychinus Ioſeph	8 Achates Gad

Coloniæ excudebat Hero Alopecius Anno. 1539.

MARBODE'S TREATISE ON GEMS: 1539 BOOK PLATE

This illustrates the Breastplate of Aaron and gemstones associated with each tribe of Israel. The revolutionary poem contained in this volume would influence generations of writers and poets.

3

APPENDIX *Biblical Baubles*

Most gems were thought to have strong Biblical connotations—being associated with the foundation stones of the New Testament (*Revelations* XXI), which in turn, can be traced to the stones of the High Priest's Breastplate as listed in *Exodus* XXVIII. Rabbinical writings state that the stones of the breastplate were also sacred to the twelve angels which guarded the Gates of Paradise.

10

MONTH	GEM	GUARDIAN ANGEL
January	Onyx	Gabriel
February	Jasper	Barchiel
March	Ruby	Malchediel
April	Topaz	Ashmodiel
May	Carbuncle	Ambriel
June	Emerald	Muriel
July	Sapphire	Verchiel
August	Diamond	Humatiel
September	Jacinth	Tsuriel
October	Agate	Barbiel
November	Amethyst	Adnachiel
December	Beryl	Humiel

12

15

APPENDIX *A Hierarchy of Metals*

The Babylonians, attributing terrestrial events to sidereal influences, found a connection between the celestial bodies and the metals that reflected their rays. Gold, they thought, corresponded with the Sun, silver with the Moon, lead with Saturn, iron with Mars, and tin with Jupiter. The Greeks also believed there was a correspondence between the planets and metals. This belief found a place in the Mithraic mysteries. In the course of Mithraic initiation, the soul ascends a ladder of seven steps, the first of which is lead, the second of tin, the third of bronze, the fourth of iron, the fifth of tempered alloy, the sixth of silver, and the seventh of gold, corresponding to the planets Saturn, Venus, Jupiter, Mercury, Mars, the Moon and the Sun.

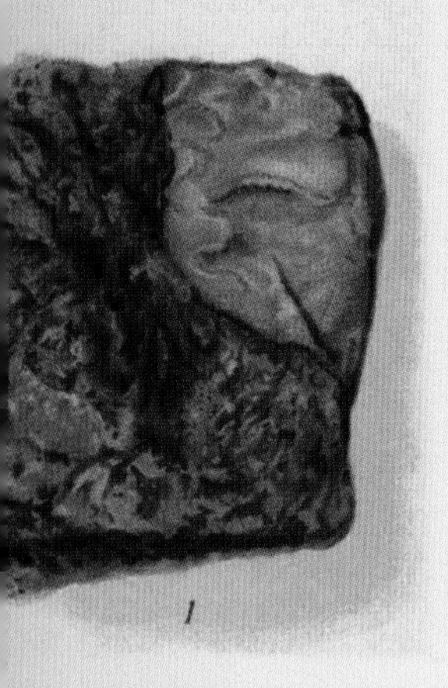

APPENDIX *A Hierarchy of Metals*

The belief persisted well into the XIV century as evident in lines 272-276 of Geoffrey Chaucer's *Conon's Yeoman's Tale* (ca. 1374).

The bodyes sevene eek, lo! hem heere anoon:
Sol gold is, and Luna silver we threpe,
Mars iren, Mercurie quyksilver we clepe,
Saturnus leed, and Juppiter is tyn,
And Venus coper, by my fader kyn!

The bodies seven look! They are here.
The Sun is gold, and the Moon silver we affirm,
Mars is iron, Mercury quicksilver we call,
Saturn is lead and Jupiter is tin,
And Venus is copper by my father's kin!

Dante follows the lead of the ancients by describing the ascent of the soul as progressing from grade to grade.

Questi organi del mondo così vanno,
come tu vedi omai, di grado in grado,
che di sù prendono e di sotto fanno.

Thus do these organs of the world proceed,
As thou perceivest now, from grade to grade;
Since from above they take, and act beneath.
(Paradiso, II, 121–123)

Symbolic of man's upward journey toward the divine, Dante refers to a golden ladder:

> *di color d'oro in che raggio tra luce*
> *vid'io uno scaleo eretto in suso*
> *tanto, che nol seguiva la mia luce.*
>
> Coloured like gold, on which the sunshine gleams,
> A stairway I beheld to such a height
> Uplifted, that mine eye pursued it not.
>
> *(Paradiso, XXI, 28–30)*

This association with the spiritual journey still resonates in modern culture with Led Zeppelin's *Stairway to Heaven* being one of the most popular rock tunes in history. The song writers were guitarist Jimmy Page and vocalist Robert Plant who recalls having been inspired by a book on mysticism when he wrote the lyrics.

Magic of Amber

IN-DEPTH

Amber has been highly prized for its magical and medicinal properties since Neolithic times. It is not actually a stone or a mineral, but an organic fossilized resin exuded by coniferous and flowering trees.

The most abundant amber deposits are found in the Baltic and date to about 130 million years ago. In the Dolomites of northern Italy, far older deposits have been discovered. There, arthropods have been found among the inclusions in the 230-million-year-old amber, making them the oldest-known specimens of their type.

Warm to the touch, amber is piezoelectric—picking up an electrostatic charge when rubbed—a property recorded by Thales of Miletus in the VII century B.C. Adding to its mystery, the Greeks noted that charged amber buttons could attract light objects such as hair. They also saw that if they rubbed the amber long enough, it would become charged and an electric spark could be emitted on contact with an uncharged conductive object. The Greeks were so taken with the special properties intrinsic to amber that they coined words to describe them. The word piezoelectric is a combination of two Greek words, *θλιβω*, which means "to press," and *ηλεκτον*, which means "amber." In fact the Greek word for amber is ēlektron, *ηλεκτον* (formed by the sun) and was adopted into modern language when electrons and electricity were discovered.

Needing to explain the origin of this

Magic of Amber IN-DEPTH

mysterious jewel, the Greeks believed that amber was formed when Phaeton, the son of Helios, the sun god, was struck by lightning. His sisters, overcome with grief, were turned into poplar trees and their tears were drops of amber.

Both Tacitus and Pliny the Elder observed that amber often contained insects and believed it was a juice or "marrow discharged by trees belonging to the pine genus" and from that, renamed it *succinum* or sap. The mineralogical name of amber follows the Latin root and is succinite. Not all amber is created equal. The secret is in the types of trees which produced it. Certain ambers are green, blue, red or brown and the chemical composition of the amber varies greatly. Sicily has its own particular amber called simetite which is dated to about 60 million years ago.

Hippocrates (460-377 B.C.) described the therapeutic properties and methods of application of Baltic amber that were used by healers until the Middle Ages. Records from medieval herbalists show recipes containing terms *Oleum succini* (amber oil), *Balsamum succini* (amber balsam), *Extractum succini* (amber extract) for treatment of various ailments. Wearing the beads themselves could also be useful as it was believed that the magic force of this yellow stone could absorb unhealthy yellowness of the skin.

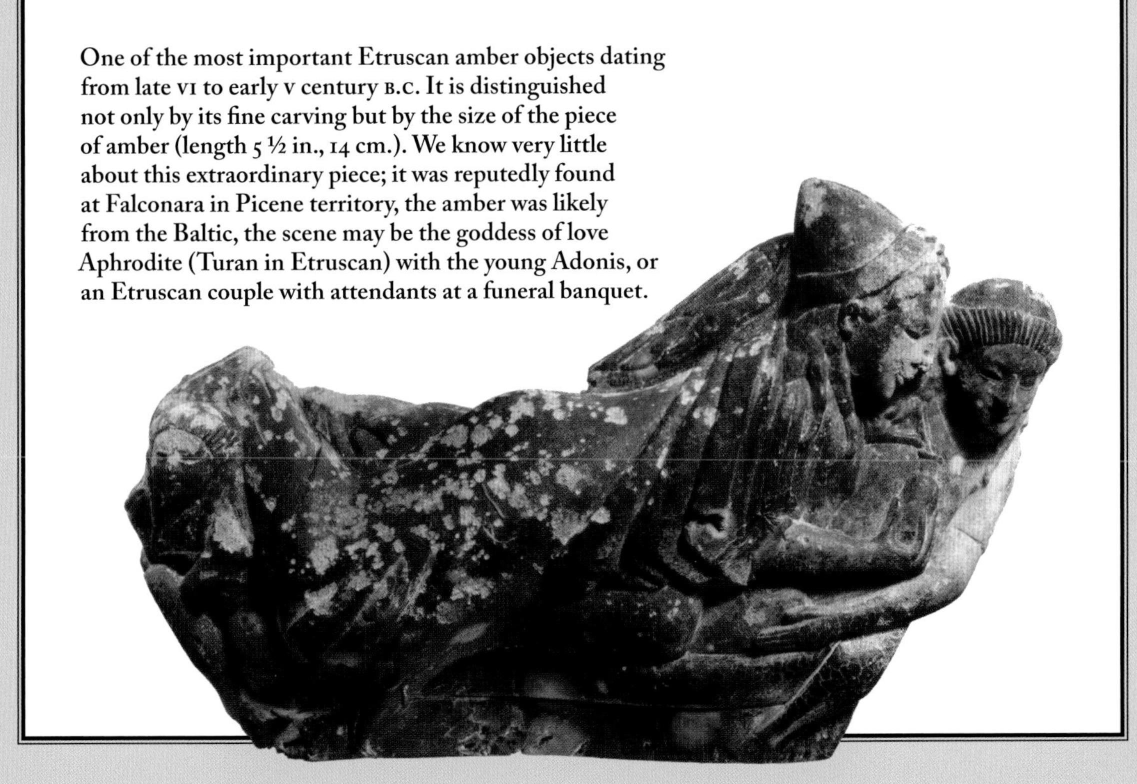

One of the most important Etruscan amber objects dating from late VI to early V century B.C. It is distinguished not only by its fine carving but by the size of the piece of amber (length 5 ½ in., 14 cm.). We know very little about this extraordinary piece; it was reputedly found at Falconara in Picene territory, the amber was likely from the Baltic, the scene may be the goddess of love Aphrodite (Turan in Etruscan) with the young Adonis, or an Etruscan couple with attendants at a funeral banquet.

IN-DEPTH | ***Magic of Amber***

In ancient Rome, Baltic amber was used as both a medicine and as a protection against disease. Pliny writes that Calistratus, a famous Athenian orator of the IV century B.C., praised amber for its powers to protect from madness. Other writers noted that powdered amber mixed with honey cured throat, ear and eye diseases and taken with water cured stomach illnesses. Diodorus Siculus mentions it being used in mourning rituals where it was dedicated to divinities and used in funeral and burial ceremonies.

Given its magical and medicinal properties, it should come as no surprise that amber has long been considered an ideal medium for the making of jewelry, talismans and small, portable works of art. Some of the earliest amber objects have been found in Italy. From the X century B.C. a major center for amber trading and craftsmen was established on the Adriatic coast at Frattesina, a convenient stopping point for merchants bringing the material from the Baltic into Italy via the Amber Route. As the demand for amber increased significantly during the VIII-VII century B.C., new trade routes were established, allowing the rest of Italy to have access to the prized material. Among the most beautiful pieces that have been found are those from tombs in Melfi and Latronico in Basilicata dating to the VII-V centuries B.C. and include pendants in the form of winged female sphinxes, which were thought to aid the deceased in the journey between this world and the next. Amber was very precious to the Etruscans, who, realizing its piezoelectric properties, considered it a magic stone. Archeologists have found spinning tools made of bronze with strings of amber beads, which, with their electrostatic properties attracted the fibers of wool, flax or hemp.

While the beauty and physical properties made amber prized as a magical stone, none of the healing effects attributed to it could be substantiated. This is where modern science solved the mystery. In the 1930s, European biochemists discovered succinic acid, which is a dicarbocylic acid ($C_4H_6O_4$), an organic compound that occurs in all living creatures. Succinic acid is a vital component of the Krebs cycle, which regulates our metabolism. It is within this cycle that carbohydrates, fats and proteins are transformed into energy. Additionally, the compound is a powerful antioxidant, bolsters the immune system, improves concentration and reduces stress.

So what does this have to do with amber? Well, one of the constituents of the fossilized amber resin is succinic acid. Not all ambers possess it; the secret is in the types of trees in which it was produced. The Baltic ambers were formed from ancient conifers which possessed abundant amounts of naturally occurring succinic acid in the resin.

Pliny the Younger noted that Roman peasant women wore amber medallions not only as adornments, but also as a remedy for "swollen glands and sore throat and palate."

FLORA AND FAUNA
Leonardo saw fossils of plants and animals on mountain tops, causing him to call into question the biblical notion that they were deposited during the Great Flood.

own; being in a constant state of transformation. Objective and scientific, Leonardo's critical observations and detailed geological knowledge are evident in his many well-known paintings, sketches, and private writings on science. One notable example is the *Codex Leicester*, a collection of 18 folios, folded in half and written on both sides, penned by the artist circa 1508–1510 and acquired by Microsoft founder Bill Gates in 1994.

With regard to fossils, Leonardo challenged many of the prevailing views of his day, including that of the Greek philosopher Aristotle (384–322 B.C.), who had written that fossil seashells were produced by the influence of the stars—an idea that would persist in scientific circles well into the XVIII century. The artist stated that this could not be possible since fossils of both flora and fauna occur at various elevations and depths within the rock. Leonardo dismissed other naïve notions—that fossils were "jokes of nature created by God" to confuse mankind or that they had been created from "rock juice" or "seminal vapors," the latter a germy mixture existing deep within the Earth. Perhaps more important, he refused to accept the biblical notion that shells had been deposited by the Great Flood, noting that, "*If the shells had been in the turbid water of a deluge, they would be found mixed up and separated one from another, amid the mud, and not in regular rows in layers as we see them in our own times.*"

The revolutionary nature of Leonardo's observations cannot be underestimated as his comments on the deposition of fossils not only questioned the facts associated with the Great Flood, but threatened the Church's official position that the Earth was a mere 6,000 years old.

Leonardo describes the formation of stones as well. "*When nature is on the point of creating stones, it produces a kind of sticky paste, which, as it dries, forms itself into a solid mass together with whatever it has enclosed there, which however, it does not change into stone but preserves within itself in the form in which it has found them. This is why leaves are found whole within rocks that are formed at the bases of mountains, together with different kinds of things, just as they have been left there by floods from the rivers that have occurred in the autumn seasons; and the mud caused by successive inundations has covered them over, and then this mud grows into one mass together with the aforesaid paste, and becomes changed into successive layers of stone, which correspond with the layers of mud.*"

NO STONE LEFT UNSKETCHED

In this marvelous design, bedded standstone is visible at the top of a cliff. Triangular shaped scree, which has flaked off the bedrock (possibly basalt) is at the bottom and boulders are seen beneath the stream's rippling water.

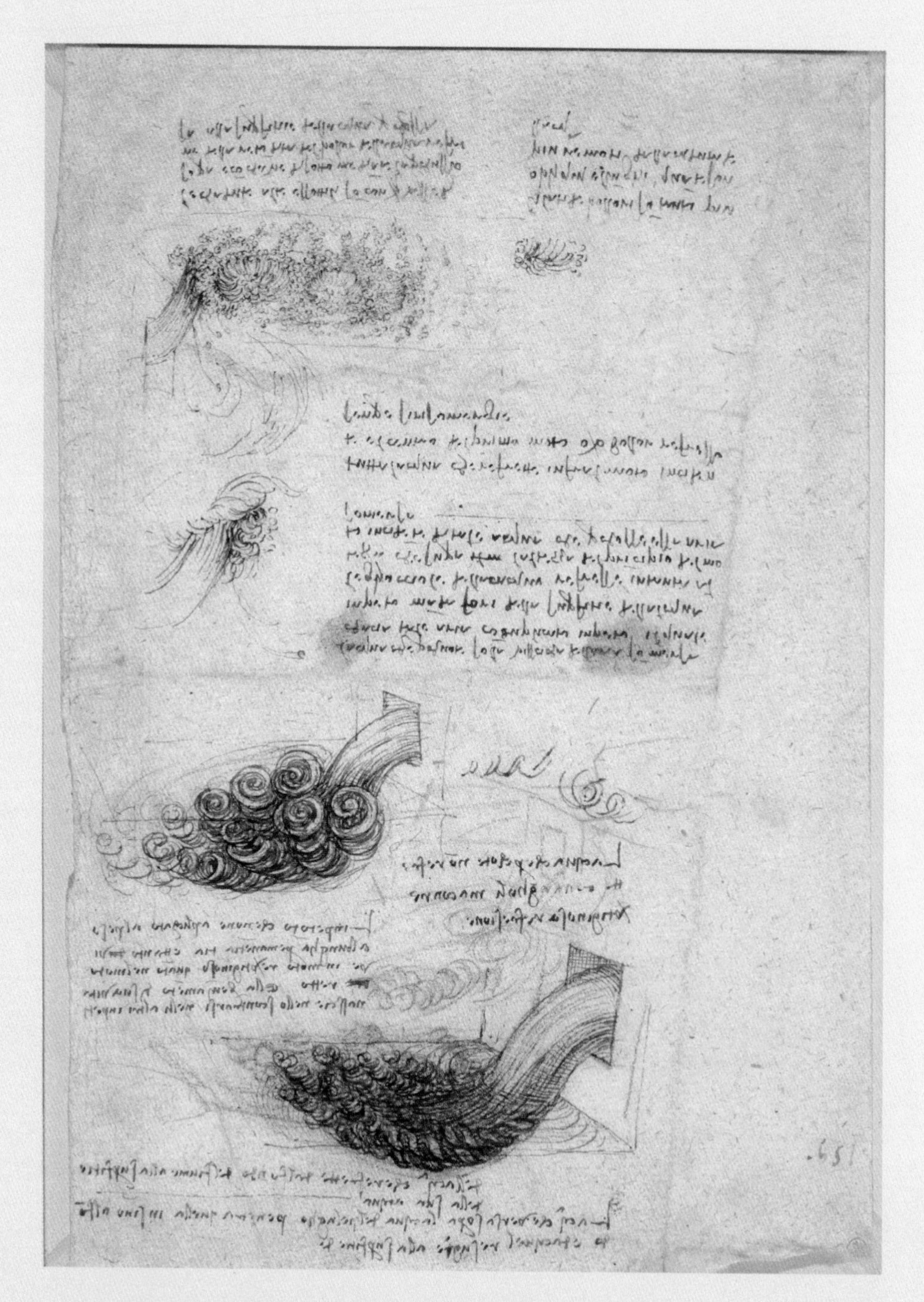

LEONARDO: THE FIRST MAN IN HISTORY TO HAVE TACKLED HYDRAULIC THEORY

Leonardo reviewed the hydraulic devices designed by Heron of Alexandria (I century B.C.). He then made painstaking notes on the movement of water and invented many devices to control, divert and measure fluids.

So accurate in its content, this passage could be taken from a modern geology textbook as Leonardo cogently described the geologic processes of deposition, fossil preservation, fossil mold formation and stratification.

In a remarkable observation, he speculated that land actually rose up out of the sea. Once it emerged, rivers and runoff eroded the terrain and moved fossils from one place to another. Commenting on the appearance of fossils on mountaintops, Leonardo wrote the following: *"The shells of oysters and other similar creatures that are born in the mud of the sea testify to us of the change in the earth round the center of our elements. This is proven as follows: the mighty rivers always flow turbid because the earth is stirred up in them through the friction of their waters upon their bed and against their banks. This process of destruction uncovers the tops of ridges formed by the layers of these shells, which are embedded in the mud of the sea where they were born when the salt waters covered them, and these same ridges were from time to*

WATER STUDIES

In his capacity as engineer, Leonardo observed water courses with the intent of building dams, canals or even diverting rivers (mainly for political and economic reasons). By studying flow, topography and meteorological conditions, he developed many ideas and instruments which would be realized centuries later.

time covered over by varying thicknesses of mud that had been brought down to the sea by the rivers in floods of varying magnitude. In this way these shells remained walled up and dead beneath this mud, which became raised to such a height that the bed of the sea emerged into the air...and now these beds are so great a height that they have become hills or lofty mountains, and the rivers that wear away at the sides of these mountains lay bare the strata of the shells."

Leonardo's commentary was some 400 years ahead of its time, for such observations would not be made again until the late XIX century when they were formalized into the specialized fields of geological studies: stratigraphy and paleontology.

Fascinated by the movement of water and its properties, Leonardo was the first to undertake a theoretical and systematic approach to what would in time become the field of fluid dynamics. He carried out a series of hydrological studies at the request of Cesare Borgia, who asked the artist to assess the potential impact of damming the Arno in the summer of 1502. More than a dozen folios within the *Codex Leicester* are dedicated to his observations on the movement of water—currents, eddies and waves—and the design of all manner of devices to channel, monitor and/or harness its power.

Engineers studying his water illustrations have commented that Leonardo must have had the ability to remember the spilt-second trajectory of droplets suspended for a moment in air because the drawings depict the exact path which these drops would have followed mathematically. In addition to his sketches, he also wrote about seeking a methodology for measuring fluids, which, not having a well-defined shape, proved impossible to quantify. Even the Greeks, who could measure solids, failed to master the measurement of liquids.

Leonardo noted that water had the potential to cause cataclysms and destroy cities, which he captured in his so-called *Deluge Series*,

WONDROUS WAVES
Leonardo correctly guessed the laws of wave motion in 1504. They were not formulated until 1673 by the Dutch physicist Christian Huygens.

[illegible]

a group of designs executed ca. 1515 and now in the collection of Windsor Castle.

Likening the Earth to the human body, Leonardo wrote: "*We may say that the Earth has a spirit of growth, and that its flesh is the soil; its bones are the successive strata of the rocks that form the mountains; its cartilage is the tufa stone; its blood, the springs of its waters. The lake of blood that lies about its head is the ocean, its breathing is by the increase and decrease of the blood in its pulses, and even so in the Earth is the ebb and flow of the sea. And the vital heat of the world is fire, which is spread throughout the Earth; and the dwelling place of its creative spirit is in the fires, which in diverse parts of the Earth are breathed out in baths and sulfur mines, and in [the volcanoes of] Vulcanus and Mongibello in Sicily and in many other places.*"

Sensitive to the portrayal of the natural world, Leonardo objected to the Renaissance practice of using vague landscapes, whose only purpose was to act as a backdrop for the more important human figures. He mentions in particular the "*very bad landscapes*" of Botticelli, adding that "*The painter is not well-rounded who does not have an equally keen interest in all things within the compass of painting.*"

In his *Treatise on Painting* (*Codex Urbinas),* Leonardo describes a painter's powers: "*The painter can call into being the essences of animals of all kinds, of plants, fruits, landscapes, rolling plains, crumbling mountains...places sweet and delightful with meadows of many colored flowers bent by the gentle motions of the wind, which turns back to look at them as it floats on.*" For Leonardo, art and science were one and inseparable. Such extraordinary powers of observation are his trademark and as such, they hold the key to determining the authenticity of his work. ❁

Leonardo's Post-Diluvian World

In a collection of eleven drawings known as the *Deluge Series*, now in the collection of Windsor Library, Leonardo da Vinci presents his cogent observations on the nature of flowing water and its impact on the landscape. Having worked as an engineer and hydrographer well before such professions would come to exist, Leonardo mapped numerous waterways and designed canals. In the process, he gained detailed knowledge about water and its motion in waterfalls, rapids, currents and torrents. He documents the storms he witnessed in the mountains and the resulting landslides and floods which took a catastrophic toll on the surrounding Alpine villages. What is sensational about these drawings is the accuracy with which the artist depicts the elements—the swirling waves, the billowing winds and the precise geometric form of the rocks as they split off the mountain. The two presented here are cases in point.

THE ARNO RIVER AND ITS WATERSHEDS
Note the extraordinary aerial prospective, as if Leonardo were in an airplane. The river, with its vein-like tributaries and intricate drainage patterns curves its way through the mountains near Pisa down to the marshlands of the Chiana valley. He drew this map during his service as engineer to Cesare Borgia circa 1502.

Town at Center of Vortex

(RL 12378) CIRCA 1515, BLACK CHALK

In this dramatic, forceful and beautiful drawing, Leonardo combines the power of earth, atmosphere and water. Their convergence results in a catastrophe, where the earth is subject to powerful forces which cause massive landslides.

During Leonardo's lifetime, large rock-slides in the Alps near Bellinzona were recorded in 1513 and in 1515. Having witnessed avalanches caused by storms, Leonardo states in his notebook "*a mountain fell seven miles across a valley and closed it up and made a lake.*" Such occurrences are still common in the Alps where atmospheric conditions are conducive to heavy rainfall which scours the terrain, often with disastrous results.

The drawing is deceptive as it demands close inspection. From a distance, one is not sure if it is a macro or micro view. Is that a city at the bottom, or details of a rock formation with a square joint pattern?

The combination of the swirling water

and falling rocks shows the integration of nature as these forces meld into one destructive element. Kenneth Clark describes the Deluge drawings as follows: "*Leonardo's scientific knowledge of nature, and his even more extraordinary intuitions as to the hidden potentialities of matter, have enabled him to pass into a different world from the old medieval Apocalypse with its confused oriental symbolism; and to arrive at a vision of destruction in which symbol and reality seem to be at one.*"

Water Breaking Through a Mountain Pass, Causing a Rockslide and Large Waves in a Lake

(RL 12380), PEN AND TWO INKS (BLACK AND YELLOW) OVER BLACK CHALK CIRCA 1515

While its title mentions a "mountain pass," the setting for this drawing is, in all likelihood, a quarry. The rocks which are falling from the steep, vertical walls in the background are not irregular boulders or sheets as might be expected in a mountain landslide, but are jointed, fractured and squared with near 90-degree angles. The flood waters rushing in at the top broke the rocks away from the wall. A quarry is also logical because there would be an accumulation of water at the bottom. Blocks crashing into the depths of such a confined space would result in large waves splashing upward. Clearly, Leonardo studied the displacement of water when an object was thrown into it.

The resulting turmoil is neither random nor chaotic but one of strong, yet ordered movement that reflects the motion, force and gravity of disparate elements as they come in contact with each other. Leonardo, as a scientist, knew how the water would react to displacement and he did not leave one pen stroke to chance. Within the drawing, one can follow the path of each wave and rock, knowing full well that each follows a trajectory governed by physics. Today, scientists describe movement and displacement in terms of angles of impact, waves, torque and speed—terminology that lay centuries beyond Leonardo's time. Yet da Vinci was also an artist, so he portrayed the waves like curled locks on a woman's head and the falling rocks resemble a stack of children's building blocks slowly collapsing. They fall in a curvilinear fashion, following the sinuous arc of the wave. Thus, a cohesiveness of form is found in a picture which at first glance appears to be violent, yet due to Leonardo's mastery, when viewed as a whole, manifests a sublime delicacy.

These designs have more secrets to reveal, as they are also a showcase for meteorological phenomena, as Leonardo accurately portrayed the flow of invisible air and noted, "*that in all cases the motion of water conforms to that of air.*" What he depicted would be discovered by meteorologists in the 1970s and called "vortex flow patterns" which are rotating masses of air or water spinning about an imaginary axis. These rotating flow patterns are similar to hurricane configurations which are so familiar to us today. ❋

APPENDIX *The Earth's Paint Box*

Man has sought to express himself through art since prehistoric time. But in order to produce creative works, materials were needed—the more colorful the better. Early man used what was at hand: plants and insects, but most of all, minerals. The earth provided a marvelous free paint box of pigments which were not only vibrant but durable. Ancient cave paintings, brick red in color, were made with ochres, an iron-oxide rich soil. Chalks were also readily available, right from the ground, with red being a preferred hue. In fact, the preference for red chalk by Renaissance masters such as Leonardo, Michelangelo and Raphael demonstrates an appreciation for this "poor man's" material which has remained popular for millennia.

The colored paints that can be bought today, mixed to sheer perfection, are a product of very recent times. Until about 100 years ago, painters still relied almost exclusively on natural products. This reliance made their work more difficult, more costly and at times fatal. In fact, Caravaggio may have died and Goya and Van Gogh may have suffered various maladies stemming from their use of lead paint. A green pigment which contained arsenic may have also been the cause of Cézanne's diabetes and Monet's blindness.

Let's look for a moment at the situation faced by artists. Minerals for paints had to be obtained, often imported. If an artisan received a commission which required a costly raw material he would sometimes have to finance the purchase in advance with apprentices often having to be sent long distances to obtain the precious substance. The products also had to be fresh otherwise they would lose color quality. Grinding was an art as well, as each material required a specific, unique consistency. The pigments were mixed either with egg white or oil depending on the final use and these procedures were guarded much like state secrets. An artist's reputation was based not only on his technique but on his use of color. Indeed, a series of notes by Leonardo da Vinci dated to about 1508 show that he was busily experimenting with mixtures that would imitate the color and design of semi-precious stones. When we think of the craftsmanship involved in layering color upon color to arrive at the one actually needed, it is astonishing that such sublime works were produced at all.

Some of the most common minerals used in the artist's palette since the dawn of time are:

BLUE

Lazurite-a tectosilicate mineral composed of sodium, calcium and aluminum among many other components, the blue color comes from an anion of sulfur (S_3-).

Lapis lazuli also known as *ultramarine*-most complicated chemical formula of the blue pigments, contains lazurite and the S_3^- anion

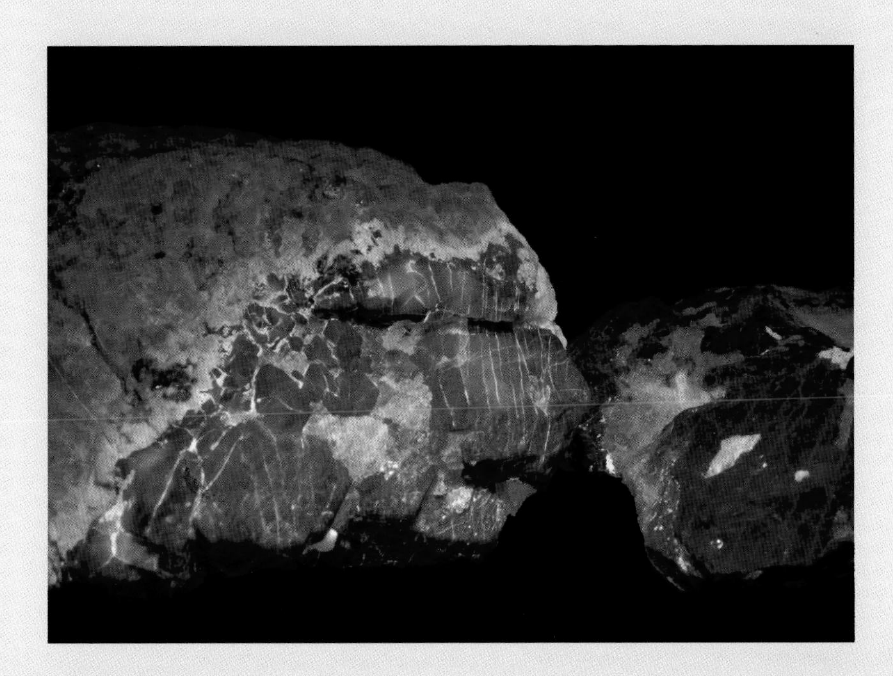

for the rich cobalt blue color. Its name derives from "beyond the sea" as it was imported from Afghanistan and was once more expensive than gold. No other material could give the same color, so substitution was not possible.

Azurite-copper carbonate, the most important blue pigment of the Middle Ages through the Renaissance.

GREEN

Verdigris-copper acetate, was the most vibrant green available until the XIX century, its color was not stable and it was toxic. It was often mixed with arsenic trioxide to form a color called *Paris Green.*

Green earth-a hydrosilicate (celadonite and glauconite), it was used since antiquity, its color was stable.

Malachite-copper carbonate, oldest known green pigment.

YELLOW

Naples yellow-lead antimonite, it was a yellow ceramic glaze in ancient Babylon and Assyria. It was preferred by the Old Masters because of its pale warmth, its use can be traced back to about 1400 B.C.

Orpiment-arsenic sulfide, one of the only clear, bright yellow pigments available to artists until the XIX century, highly toxic.

Lead-tin yellow-lead tin oxide discovered in the XIII century and used most common in the XV-XVII century.

ORANGE

Realgar-arsenic sulfide, only pure orange pigment available until recently, highly toxic.

RED

Cinnabar-mercury sulfide, the Romans combined it with sulfur to form an expensive red paint also known as vermillion. It was mined at Almaden, Spain, but also comes from China, highly toxic.

Hematite-from the Latin for red or blood, the iron in this mineral accounts for the red color.

RED/YELLOW/BROWN

Ochre-an earth pigment rich in iron oxide, limonite gives a yellow color; hematite, red. The color varies according to the amount of iron.

Chalk-ready for immediate use, comes in various colors.

WHITE

Lead white-Greeks developed it, taking the mineral cerussite, a lead carbonate, and mixing it with metallic lead and vinegar. It was the only white used until XIX century, highly toxic.

Lime white-chalk plus Bianco San Giovanni (calcium carbonate plus calcium hydroxide).

BLACK

Carbon black-burned bone, ivory, wood or plant material.

V. Leonardo's Geology: A Tale of Two Paintings

Leonardo da Vinci was a consummate observer of nature. His scientific curiosity led him to depict natural objects not only exquisitely, but accurately as well. His sketches and drawings serve as a record of the geological formations he saw in his travels. Most of his early years were spent in Florence, but it was when he went to Milan to work for Ludovico Sforza in 1482, that he became fascinated with Alpine geology. He spent considerable time in the mountains observing the structure of rock formations, the presence of fossils imbedded in stone, plants and atmospheric phenomena. His observations on geology, hydrology and the effects of water and air on the earth are still visible in his note book, the *Codex Leicester,* owned by Bill Gates. Because of his artistic prowess, he was able to incorporate into his paintings and drawings precise geological formations which, at the time had not been named, but which are readily identifiable to a modern geologist.

Viewed from a geological perspective, all of Leonardo's paintings and drawings reveal a remarkable fidelity to nature. The *Virgin of the Rocks* in the National Gallery in London (1495-1508), attributed to him, displays no such fidelity. If we compare it to the *Virgin of the Rocks* in the Louvre in Paris (1483-86) whose geological accuracy is astounding, we cannot help questioning whether Leonardo painted the background in the National Gallery work.

Over the centuries, various arguments have called into question the attribution of the National Gallery painting to Leonardo. Scholars have analyzed the brush strokes, undertaken document searches and tried to prove definitively that Leonardo produced the National Gallery version. However, there have always been doubts, naysayers and many unanswered questions concerning its authenticity.

The fact that the attribution of the work has been the subject of

PICTURE PERFECT GEOLOGY
This detail from the Louvre shows Leonardo's mastery of art and geology. He paints a complex landscape with such accuracy that each rock formation can be identified. To the right of the Virgin's head is weathered sandstone, above it is a contact surface with a strata (sill) of diabase (note the change in texture between the two) and above that is spheroidal sandstone.

such controversy throughout history suggests that new diagnostic means of determining authenticity is in order. A comparison of the representations of geological configurations in the two paintings offers such a means. It seems unlikely that the same person could have portrayed rock formations so accurately in the Louvre work and so incongruously in the National Gallery painting.

The *Virgin of the Rocks* in the Louvre is a geologic tour-de-force because of the subtlety with which Leonardo represents a complicated geological assemblage. At the top of the grotto are rounded (spherically weathered) mounds of sandstone, a sedimentary rock. Above the Virgin's head is a rock which extends upward with distinct vertical relief. This is diabase, an igneous rock that was injected as a molten liquid and spread over the sandstone, forming a band (or a sill) several feet high. The rock contracted as it cooled, forming vertical (columnar) joints. Directly above the Virgin's head is a

LOUVRE VERSION
This painting was executed with such precision that geologists today can identify many marvelous features. To portray these details meticulously required not only a talented artist, but a person who recognized, understood and appreciated geology.

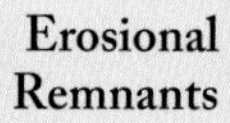

NATIONAL GALLERY, LONDON PAINTING
The rocks depicted in this version are not realistic. The artist lacked both technique and an appreciation for the geology being portrayed. The rocks are all angular and block-like with no distinctive texture. The steep drop-off in the foreground gives the impression that infant is about to fall into an abyss.

horizontal crack in the rocks called a basal or bottom contact. This is the seam between the diabase above and the sandstone below. The column of diabase extends upward until it meets another horizontal contact surface and the rock formation changes to sandstone at the top of the grotto.

The rocks that extend from below the basal contact line near the Virgin's head down to the foreground are sandstone, like those at the top of the grotto. The texture and rounded weathering pattern of the sandstone are the same below the basal contact as they are

BOTANICAL BRILLIANCE **Leonardo appreciated the beauty found in the natural world as evidenced by his accurate and stunningly beautiful depiction of the undulating leaves of this Star of Bethlehem and other plants. Interestingly, this sketch is dated ca. 1506-08, coincident with the National Gallery painting, so it seems highly unlikely Leonardo painted the inaccurate plants found in that version.**

..and to me there is a surprising—if not shocking—difference in the plants in the landscape. In the French painting the plants are beautifully rendered, with the detail one expects from the great botanical artist that Leonardo was: an Iris, Polemonium and Aquilegia are clearly recognisable. Replacing the Iris in the London version is a clump of apparent Narcissus tazetta - but it is no normal daffodil. The flowers are good enough, but they arise on bracteose scapes, from a clump of plantain-like leaves. Next to this are two completely fantastical plants that cannot be identified, and there are other oddities elsewhere in the landscape. It seems that here Leonardo, the inveterate doodler and inventor, has invented his own flowers for Paradise, and in the evolution of the painting has translated the scene from an earthly to a heavenly location. It is nice to think that he envisaged daffodils there."

—*John Grimshaw, horticulturalist*

BOTANICAL DIFFERENCES

The botany in the Louvre version (above) is perfect, depicting plants which would have thrived in a moist, dark grotto. The ones in the London version (below) are inaccurate botanically, some not existing in nature and others portraying flowers with the wrong number of petals.

THE VIRGIN AND ST. ANNE

Leonardo continued to experiment with more detailed rock formations. In this 1510 work in the Louvre, he developed new paints and techniques to be able to depict the agate and chalcedony pebbles (BELOW) in the most realistic manner possible.

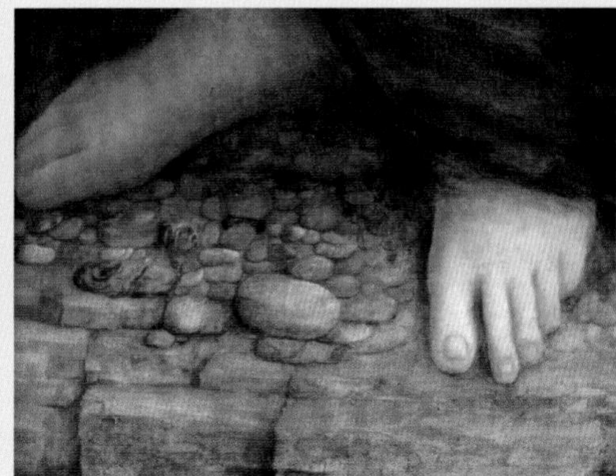

SUBLIME SANDSTONE

When on his frequent trips into the Alps, Leonardo would always sketch the natural formations he saw. These sketches are the reason why his paintings and commentary on natural processes are so accurate. Here, bedded sandstone has been subject to erosion, causing the upper layers to become detached from the bottom ones and wearing down their edges to a rounded form.

at the top of the grotto. In the foreground, the sandstone has not been heavily weathered and has therefore retained its highly defined horizontally layered (or bedded) structure. The diabase sill at the center of the formation is harder and less prone to erosion, hence its sharp edges and vertical relief. Leonardo is able to capture this contrast first, by having an understanding of how the rocks actually look and then representing their appearance realistically through his use of light and color. Leonardo's use of *sfumato*, a shading technique he mastered, imparts the feeling of a moist, musty grotto.

LOUVRE: PERFECT PLANTS
The iris, probably Iris germanica, is in the foreground. The grouping looks like an assemblage from a Mediterranean climatic region. The compound leaves on the branches at the upper left, in particular, look like leathery, evergreen sclerophyllis.

The jagged rocks rising from a blue-gray mist in the background are remnants of erosional processes that stripped away the overlying softer rock and left the remaining harder rock intact. These formations have been subtly yet accurately depicted, consistent with Leonardo's unwavering commitment to geological realism.

Especially intriguing is Leonardo's placement of vegetation in the picture—not simply to achieve an harmonious aesthetic assemblage—but rather, to place the plants in areas where they grew naturally. At the top of the grotto, the sandstone would have decomposed sufficiently to allow roots to take hold. This is also true for those growing in the foreground and in the background. No plants are growing out of the diabase, however, since it is too hard and resistant to erosion to provide a suitable habitat for plant growth.

An observer with some knowledge of geology would find that the rock formations represented in the National Gallery work do not correspond to nature, as do most of Leonardo's drawings and paintings. All we know about da Vinci suggests that he had too much respect for natural beauty to portray it inaccurately. The rocks in the National Gallery version miss the point geologically. Looking at the painting, above the Virgin's head, there is no change in the texture of the rocks to indicate the presence of the diabase sill. The vertical joint patterns continue upward without interruption. The type of rock remains constant, in comparison to the changes in rock form in the Louvre work. In the foreground, the rocks are not finely bedded and in fact, they are simply not identifiable. The lack of understanding on the part of the painter of the National Gallery work seems to exclude the possibility that it was Leonardo.

If we take this a step further, let's look at a time line of three of Leonardo's works: *Virgin of the Rocks* in the Louvre, painted ca. 1483-86, the National Gallery version ca. 1495-1508 and the *Virgin and St. Anne* in 1510. It seems unlikely that Leonardo changed his geologic style for just one painting—that in the National Gallery—considering that the *Virgin and St. Anne*, finished after the London painting, is a much more detailed and geologically complex picture.

In the end, Da Vinci's extraordinary knowledge provides us with an unbiased method of distinguishing his work from that of his many imitators and followers. Precise geology is an index to authenticity. It can serve as Leonardo's inimitable trademark as no other artist of his time understood geology so well.

An Ambiguous Past

The history of the two paintings has baffled art historians for years and debates and arguments have raged over the attribution of the *Virgin of the Rocks* in the National Gallery to Leonardo da Vinci. This mystery, with attendant acrimony, is still going on despite the fact that abundant testimony exists from a 25 year lawsuit involving the two paintings.

According to historical records, the Brethren of the Immaculate Conception (Confraternity), commissioned Leonardo and the brothers Ambrogio and Evangelista de Predis to create an altarpiece for their chapel, San Francesco il Grande, in Milan in 1483—a project estimated to cost some 800 lire to execute. The central feature of the altarpiece was to be a portrait of the Virgin Mary and Child surrounded by angels, representing the Immaculate Conception. Leonardo was to prepare the painting and the de Predis brothers were to complete the frame. Though da Vinci completed the piece by 1486, it remained uninstalled for some four years while the de Predis brothers worked to finish its ornate frame.

In about 1490, the de Predis brothers appealed for more money, citing that the frame alone had cost the entire amount to which the artists had originally agreed. They asked that the "oil painting of Our Lady" (*Virgin of the Rocks,* Louvre) be withdrawn

IN-DEPTH

An Ambiguous Past

EMBATTLED ALTAR
While it has not survived, this is a sketch of the altarpiece that was the subject of a 25 year legal battle involving Leonardo, the dePredis brothers and the sponsor. The lawsuit is the reason why there are two versions of the central painting of the *Virgin of the Rocks*.

from the commission as "others" had offered to purchase it, presumably for more money.

The legal wrangling went on for almost a quarter century. The exact motivation is unclear because there were many claims and counterclaims. Leonardo and the de Predis brothers wanted more money, while the Confraternity argued that the picture did not fulfill their request of a painting representing the Immaculate Conception of the Blessed Virgin and therefore called the painting "unfinished" to bolster their position in court.

Perhaps the Confraternity could not

accept Leonardo's painting because the conditions for funding the project were very specific and would not allow for variation or substitution. Ambrogio de Predis found this out when he petioned the court in 1503, as the Confraternity was not willing to accept the painting and Leonardo was unwilling to drastically alter the one he had completed (Louvre), nor paint another.

Charles Hope, expert in notarial Latin and director of the Warburg Institute in London has completed an exhaustive study of these arcane court documents and tells us what happened next. "Leonardo and the de Predis brothers had hoped to receive at least 400 lire, and the patrons initially offered only 100. In 1506 they raised this figure to 200, on the condition that the picture was finished. Had this happened, Leonardo would evidently have considered himself out of pocket, even though after 1506 he and Ambrogio did indeed receive 200 lire. Although the documents are silent on this point, it looks as if the patrons finally accepted the second version, for a reduced fee, and returned the original (Louvre) to Leonardo, who was able to recoup his full fee, including Ambrogio's share, by selling it to a third party. How and when it entered the French royal collection has been much debated. But it may well be relevant that in 1508 Milan was under French control and that Leonardo had been given a salary by Louis XII. Although he had proposed to Ambrogio that they should sell the copy and share the proceeds, it is certain that the copy remained in the church, and it is this copy that was later acquired by the National Gallery."

Once the legal matter had settled, the copy, now in the National Gallery was painted. Charles Hope states "Leonardo's own involvement, if there was any at all, is likely to have been very limited." Professor Hope echoes many art historians who have questioned the attribution of the National Gallery work to Leonardo.

In the ensuing years the artists parted company. The de Predis brothers remained in Milan, while Leonardo lived out the last years of his life as guest of the King of France, Francis I, in his chateau in the Loire Valley. Each painting also went its separate way. The Louvre version is first mentioned as part of the royal collection at Fontainebleau in 1625. The London version, which remained in the church of San Francesco il Grande until 1781, was taken to the hospital of Santa Caterina in Milan and sold in 1785 to the English painter Gavin Hamilton. It was in the collection of the Marquis of Landsdown, then the Earl of Suffolk, for nearly a century before entering the National Gallery in 1880. ❋

Immaculate or not?

The idea of the Blessed Virgin Mary's conception without sin was always fraught with controversy. Leonardo da Vinci would end up being caught up in this turmoil. In February 1476, Pope Sixtus IV, a Franciscan, permitted various dioceses, including his own in Rome, to celebrate the feast of the conception of Mary. He would go on to formalize his position a year later, when on February 28, 1477, he issued a Papal Bull called *Cum Praeexcelsa* which presented the sacred conception of the Blessed Virgin Mary without actually using the word immaculate, choosing "miraculous" instead. He offered indulgences for those who worshipped on this feast day, December 8. The Dominicans held a different position however, and refused to accept this interpretation. Certain Dominican theologians were known as Maculists, promulgating the idea that Mary had been touched by the effects of Adam's fall. They launched bitter attacks against the Franciscans, called Immaculists and they engaged in a raging theological and philosophical battle.

In 1483, a Confraternity of lay Franciscan supporters decided to commission a painting in honor of the Immaculate Conception of the Virgin Mary. They hired Leonardo da Vinci to paint the central panel of the Virgin and the de Predis brothers to carve and gild the frame. Leonardo had just arrived in Milan and was enamored with the Alpine geology he saw in the mountains north of the city. Further, the style of incorporating rocks in religious paintings was gathering increasing acceptance despite being espoused during the time of St. Francis some 250 years before. When Leonardo received his commission from the Franciscans, the contract was very specific: the painting was to have "*...mountains and detailed rock formations painted in oil, in many colors.*"

In pondering the request to convey the concept of the Immaculate Conception, Leonardo's challenge lay in how to present an abstract, controversial idea in an understandable manner. He remembered the striking geological formations he saw in his forays into the Alps, including a mysterious cavern. In a stirring passage in the *Codex Arundel* (155r) he describes his feelings.

"*Wishing to see the great abundance of varied and strange forms made by the art of nature,...I arrived at the entrance to a great cavern, before which...bent double and steadying my weary hand on my knee, with my right hand I shaded and lowered and half-closed eyes, frequently leaning to one side and the other to see whether anything could be discerned inside, which prohibited me by the great darkness there. And (when I had) been (there) a while, there suddenly arose in me both fear and desire: fear for the menacing and obscure cavern; desire to see whether something miraculous might lie inside.*"

This mystical, almost miraculous geological experience, which harkens back to the Classical view of caverns, gave Leonardo the inspiration as to how to portray the Immaculate Conception. And so, his choice of a grotto was not arbitrary. Aside from the fact that the Franciscans had already given him specific instructions to go full-out in the depiction of the rock formations he found so fascinating, he also knew that symbolically, the grotto had always represented the feminine dimension. Whether it was the idea of fertility or the womb being the organ of reproduction, the grotto represented the female symbol of conception, birth and life. His resulting *Virgin of the Rocks* (Louvre) was a triumph of symbolism, diplomacy and natural science all wrapped into one perfect painting.

VIRGIN AND ST. ANNE **Leonardo's working sketch, known as the *Burlington House Cartoon* is a masterpiece in its own right.**

The noblest pleasure is the joy of understanding.

—Leonardo da Vinci

Learning from a Master

It is clear from the work of da Vinci's students that he impressed upon them the importance of having an exacting eye when it came to depicting natural elements such as geological features, botanical specimens and bodies of water. Giovanni Antonio Boltraffio and Marco d'Oggiono apprenticed with Leonardo during his stay in Milan (1482–1499). Although their technique falls short of that of their teacher, it is clear that they emulated him in the rendering of the landscape in their *Resurrection of Christ with St. Leonardo of Noblac* and *St. Lucia of Syracuse* (ca.1491).

The *Resurrection* resembles many of Leonardo's works in that it depicts a scene unfolding on a rocky foreground with a vertical mass of rock behind and a winding river in the distance. The landscape presented in the *Resurrection* is reminiscent of one the students may have encountered along the rivers of northern Italy outside of Milan.

Christ, the central figure in the painting, emerges from a sarcophagus that appears to have been carved from fine marble and placed in cave of bedded sandstone. Saints Leonardo and Lucia, who witness the event, kneel atop matted soil which has been eroded by water. A network of shallow erosional patterns radiates from a drainage channel that runs between the two saints and toward the tomb chamber. While the drainage patterns are geologically accurate, the irregularly shaped pebbles placed within them lack any evidence of grainy weathering from erosion.

Rocks reminiscent of the spheroidally weathered sandstone in Leonardo's *Virgin of the Rocks* (Louvre) can be seen just behind St. Lucia. While there is sufficient detail in the rendering and placement to surmise that they had broken off from the formation above and tumbled to the ground only to be further weathered, their vague texture makes them look more like brown clouds ready to float away. But here again, the geology is accurate, the technique is what is lacking.

The rocks above St. Lucia's head, which appear harder and sharper, are clearly meant to represent diabase, a rock of volcanic origin which is formed by the cooling of a molten magma. As it cooled and hardened, vertical cracks or joints formed. This complicated geological feature is clearly Leonardesque and it is evident that his students put forth a valiant effort to depict it accurately. And

LEONARDO'S STUDENTS STUDIED GEOLOGY

Leonardo had few students, but he impressed on them the value of accuracy in depicting nature. Here, his students do him proud by presenting a work chock full of geologic detail. While they did not possess the artistic talent of their teacher they were able to depict geology that is correct and readily identifiable.

while it is not the most realistic representation due only to their artistic capabilities, it is readily identifiable geologically.

To the right of Christ's robe, is a horizontal crack, the contact-plane where two different types of rocks meet. The layered or bedded sandstone above this formation is quite recognizable and accurate geologically. It could be made more believable however, by a softening of the rock's blunt edges. The *Virgin of the Rocks* (Louvre) offers a fine depiction of this type of rock in the foreground.

Another horizontal plane, the basal (or bottom) contact where the diabase meets the sandstone in the foreground, is hidden just behind St. Lucia. While unseen, this juncture is implied geologically by the change in the type of rock. This is a sophisticated geologic concept and the subtlety in which it is portrayed is admirable.

Sandstone is depicted in two forms, horizontally bedded and spheriodally weathered, the result of rain and wind destroying visible traces of the original bedding structure.

In the distance, a river winds its way around massive pinnacles, perhaps of dolomite. Such pinnacles are formed when the forces of wind and water erode softer, overlying rock and sediments, leaving the harder rock behind. The pinnacles depicted in the *Resurrection* have been rendered in a way that conveys their inherent hardness and saw-toothed nature. They are painted with the blue-gray colors which Leonardo recommended as optimal for depicting background atmosphere.

As the river meanders, it leaves sediment on the inside bend, known as a point bar. Such deposits of sand and gravel generally have a gentle slope. In the *Resurrection*, however, the point bars have small peaks. The area scoured by the river, which is called a cut bank, or river cliff, is accurately portrayed. However, since cut banks are formed where the river is flowing the fastest, they are steep and unstable. Therefore, the placement of the city upon the cut bank is dangerous, as it is an area of active erosion.

The background is devoid of the sense of dimension or dramatic atmospheric conditions evident in the works of Leonardo and the hard, clean edges of the rocks depicted in the *Resurrection*—thought to be largely the work of d'Oggiono—lack the detailed textures and weathering patterns da Vinci created through the use of *sfumato*, a delicate form of shading. Despite deficiencies in technique, the overall attention to and accurate depiction of geologic detail sets Boltraffio and d'Oggiono apart from other painters working in Italy at the time and can be used as a method to identify their works as students of Leonardo. ❋

VI. Recesses of the Mind and Soul

...cave worship is older than any god or devil. It is the cult of the feminine principle—a relic of that aboriginal obsession of mankind to shelter in some Cloven Rock of Ages, in the sacred womb of Mother Earth who gives us food and receives us after death. Grotto apparitions, old and new, are but the popular explanations of this dim primordial craving and hierophants of all ages have understood the commercial value of the holy shudder which penetrates in these caverns to the heart of worshipers, attuning them to godly deeds.

—Norman Douglas, Old Calabria (1915)

Italy is filled with miraculous caverns. In fact, the whole country seems to be dotted with them, much like Swiss cheese. And perhaps in the Swiss cheese analogy lies the secret of the country's mystique: the prolific number of caverns, caves, grottoes and craters which pock mark the landscape. It is the geology of Italy and its ongoing deformation which make it chock full of apertures penetrating into the depths of the Earth. These are the recesses which, since the dawn of time, have served a multitude of purposes for man: shelter, sustenance and sanctuaries.

There are many notable ancient caves in the world, Altamira, Lascaux and Chauvet, to name a few. They are filled with art of rare beauty and from an anthropological viewpoint, give valuable testimony to the development of man. But these sites no longer host the rituals that gave life and function to the space. Some are closed and others accessible only to scholars. Everything is different in Italy. Most caverns are still open and freely used, even today. In fact, they have been handed down from generation to generation, one religion to another; as if the sacrality of the site was augmenting in the same way the rocks were accreting.

Italy's geology has made it perfect for the development of caves, caverns and grottoes. Vast expanses of limestone, the most desirable rock for large cavern formation, are present both above and below ground. Glacial and volcanic terrain can also be easily excavated by man and fashioned into caves. The extensive coastline is dotted with sea grottoes created by the incessant movement of the waves. All this, combined with the constant pushing and pulling of the continental plates provides a myriad of mechanisms for the creation of new apertures in the Earth.

Early man, arriving in Italy, had a primary need for shelter. The most obvious and ready made accommodation was a cave. Italy's abundance of them allowed communities to form, providing for the common good and safety of all. Once settlements were established,

MILLIONS OF YEARS IN THE MAKING **Left undisturbed in an underground sanctuary, nature can produce sculptures of incredible beauty.**

religion developed and caverns became the loci of sacred rituals. The awe inspiring configuration of large underground caverns with their marvelous stalactites, created by millions of years of activity, provided the proper setting. Their connection with the interior of the Earth was mysterious, mystical and profound. Often caverns seemed to have no end, inviting speculation that their terminus was in the Underworld or the womb of the Earth. Yet despite their association with the netherworld, many caverns often invoked the divine, being comparable to magnificent cathedrals; light and airy due to the soft colors of the limestone stalactites. As if nature's decorator had arranged a perfect setting for devotion.

NATURE'S ARTISTRY
The various forms of limestone: LEFT: a column will form when a stalactite and stalagmite meet, CENTER: flowstone, when streams of calcium laden waters dry, leaving sculpted rivulets of rock, RIGHT: soda straws, form when tiny droplets of calcium containing water solidify into delicate configurations.

Clefts in the Earth

The term cave refers to a natural opening, usually in rock, that is large enough for human entry, while caverns and multi-chambered cave systems tend to extend deeper in to the earth. The word cave can also refer to much smaller openings such as sea caves, rock-shelters and grottoes. Caves can form in practically any variety of rock depending on the environmental and meteorological conditions which cause erosion or deformation. Foremost among these formative processes are tectonic events such as earthquakes and volcanoes and erosion from water and wind.

The largest and most complex cave systems tend to form in landscapes rich in limestone, called karst, which is easily dissolved by the natural carbonic acid (H_2CO_3) found in rainwater and groundwater. This acidic water seeps through bedding planes, faults and joints, descending into the earth, etching it away slowly until caverns are formed. In the process, this water bonds with the calcium rich limestone to form calcium carbonate. It is in these so-called solution caverns that magnificent formations are found—stalactites, stalagmites, soda straws, columns and flowstones—created by the slow and steady precipitation of drip water rich in calcium carbonate.

Cave formation in Italy has been wrought largely by the tumultuous tectonic forces that gave rise to the peninsula. Lands east

Limestone Cavern Formation

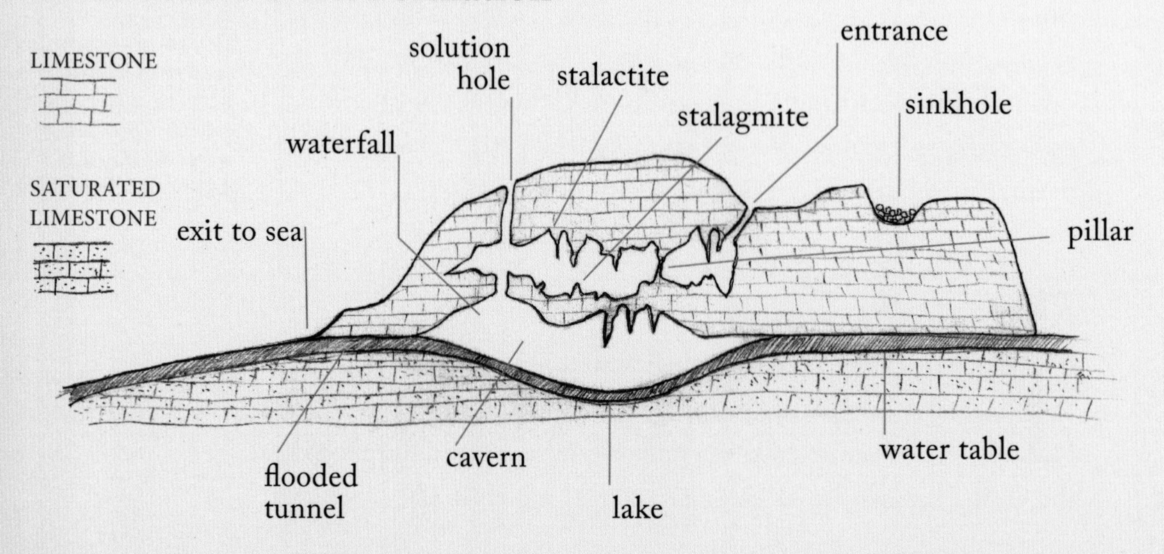

of the Apennines are largely of karst, which have been sculpted by drip water and the relentless action of the pounding surf along the coast. Lands to the west of the mountain chain are primarily of volcanic origin—rich in a variety of igneous rock and deep layers of compacted ash known as tuff. The contorted folds of metamorphic rock that dominate the landscape of northern Italy have been shaped over the ages not only by compression and uplift but by massive glaciers which scoured the land for millennia.

Despite pronounced regional variation in the overall geological make up of Italy, pockets of karst suitable for cave formation can be found just about anywhere on the peninsula. It is in these pockets of limestone that some of the country's most spectacular "show caves" can be found—such as Castellana in Apulia, Frasassi in Marche and at Toirano in Liguria.

Beyond their sheer beauty however, it is the activities that have taken place within these caves and the ancient beliefs associated with them that are fascinating. For these dark, secretive locations, which were full of exotic formations, limpid pools, moist walls and mysterious drafts provided ample evidence for immanence of the divine while low levels of carbon dioxide (CO_2) was the likely cause of visions and revelations.

Places of Transition and Rebirth

The Apulia (Puglia) region was created by limestone formed millions of years ago in a tropical sea. When the blocks of rock emerged from the water, they attached themselves to Italy, forming the famous "heel" on the "boot" of the country. The limestone, etched and eroded over millions of years, formed the caves and caverns which dot the landscape.

Not surprisingly, the earliest evidence of human occupation in Italy—and of Europe as a whole—has come from a cave site in Apulia known as Pirro Nord, located high upon the Gargano promontory in Foggia. The peninsula's first inhabitants took shelter there sometime between 1.7 and 1.3 million years ago. It took well over a million years however, for evidence of religious activity in caves to be documented. It was found just south of Pirro Nord, in the Grotta Paglicci, which provides the earliest record of rituals, dating to the Upper Paleolithic (50-10,000 years ago).

Some 12,000 to 14,000 years ago, caves came to have a close association with female fecundity and regeneration, as evidenced by the discovery of two "Venus" figurines—of clearly pregnant women—carved from bone at La Grotta delle Veneri di Parabita in southern Apulia. Complementing these objects, are the murals which abound with fertility symbols found at the nearby Grotta Romanelli.

It is in the Neolithic period (8,000-6,000 B.C.) that caves in Italy—and more specifically, grottoes that harbor mysterious water sources within them, reminiscent of the female womb and stalactites representative of the male member—become focal points for rites associated with male initiation and female fertility. These water sources include mantic springs, subterranean lakes, moisture running down cave walls and *acqua dello stillicido*, the calcite-laden drops of water which slowly drip from stalactites.

One of the most mysterious cult caverns in the region is the Grotta di Scaloria in Manfredonia. Found in 1932, the cavern was filled with carefully painted vases which were used to capture the drip water of the stalactites. Over the millennia, these vases, left in situ, were encased in limestone and permanently attached to the ground and often, stalagmites grew out of them. While archeologists have been able to roughly date these objects to circa 6,000 B.C. nothing definitive can be said about their exact use or the deities, if any, involved.

A similar instance of drip water collection is known from the

WOMEN WATER BEARERS
BELOW: Note the vases (hydria) carried by Etruscan women, v century B.C. LEFT: A female water merchant (*acquaiola*) XIX century, in Naples whose terracotta drinking cups were called *mummarelle* meaning "little breasts." The water, rich in iron, calcium and other minerals, was ingested much like a multivitamin.

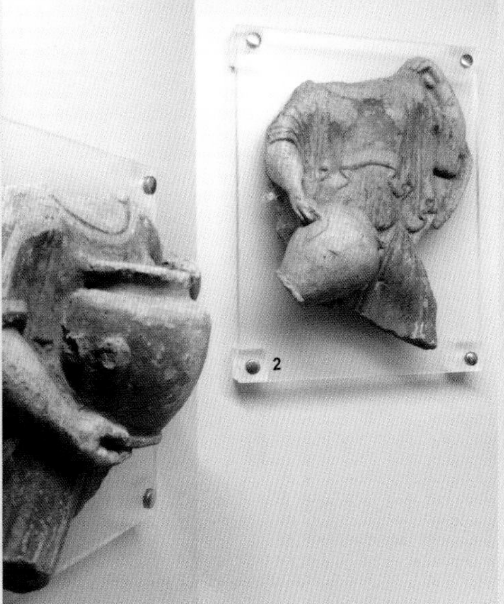

Grotta di Santa Lucia sul Monte Soratte near Rome where a large vessel was placed under a stalactite at the end of a long, difficult to access gallery. Gradually, offerings of small flasks, pigments and statues began to be deposited alongside underground lakes and within shallow pools at sites throughout Italy such as the Grotta di Pozzi della Piana in Umbria, the Grotta Verde di Alghero in Sardinia and the Grotta Zinulusa in Lecce.

The practice of collecting drip water in caves and depositing small offerings near underground water sources intensifies during

GROTTO OF SCALORIA
Within a dark cavern, votive vases were left to collect drip water. Over the millennia, the vases were calcified in-situ and stalagmites grew within them.

the Bronze Age. Within the Grotte di Pertosa, an extensive cave system etched into the foothills of the Alburni Massif 100 kilometers south of Naples, archaeologists found numerous pots and more than 300 miniature vessels which had been deposited ca. 1500 B.C. in the cave's deepest recesses. The cave itself is notable in that the river Negro runs through the system.

Given that throughout this time period—from the Neolithic through Bronze Ages, inhabitants of these regions had easy access to abundant freshwater sources to meet their daily needs, the collection of drip water and veneration of underwater pools in the most inaccessible of locations suggests that these waters were perceived as sacred and had mysterious properties associated with them.

Of Mammaries and Moonmilk

Just how such cave waters may have been used in deep antiquity comes into sharp focus during the Etruscan and Roman periods, when the cloudy, mineral-laden liquid issuing forth from the mammiform stalactites was collected for use as an emollient to ensure nursing mothers produced ample breast milk. Such caverns came to be known as milk caves (*lattaie*) with the slightly supersaturated waters they produced called mountain milk (*latte di monte*) or moonmilk (*latte di luna*). The term moonmilk is a reference to an early notion that the substance was produced when the Moon's rays flowed through rock. Among the best-known of these caves is the Grotta Sant'Angelo at Palombaro in Abruzzo, the use of which can be traced back to the Roman period. At that time, it was a sanctuary dedicated to the fertility goddess *Dea Bona* (Good Goddess). Four cisterns were cut into the rock beneath large stalactites to collect drip water—Bona's breast milk. This strong association of caves with breast milk is further underscored by Rome's founding myth of the royal twins Romulus and Remus, who were suckled in a cave by a she-wolf.

THE CAPITOLINE WOLF
This bronze sculpture of a she wolf suckling the twin infants Romulus and Remus has become the symbol of Rome. It represents the myth of the founding of the city. The sculpture itself is a mystery. Some think it is Etruscan from the v century B.C. while others feel it is from the XIII century A.D. with the figures of Romulus and Remus added in the XV century A.D. by Antonio Pollaiuolo.

So what might account for the purported healing properties of moonmilk? The secret lies in the presence of a microbe, *Macromonas bipunctata*, a Gram-negative bacterium composed partly of calcium carboxylate ($CaCO_3$) which plays a key role in metabolizing organic acids, along with a host of cyanobacteria, fungi, green algae and actinomycetes. The latter is a group of Gram-positive bacteria that are a naturally occurring antibiotic. Collectively, these agents are known to promote healing.

Calcite Catechism

Ever quick to appropriate cult practices with known efficacy, Christianity delegated the ritual use of cave waters in Italy to the purview of St. Agatha, patroness of wet nurses. Today, in Catania, St. Agatha's patron city, candles in the form of breasts can still be purchased. Yet the connection between caves and Christianity runs far deeper. One need only to look to the

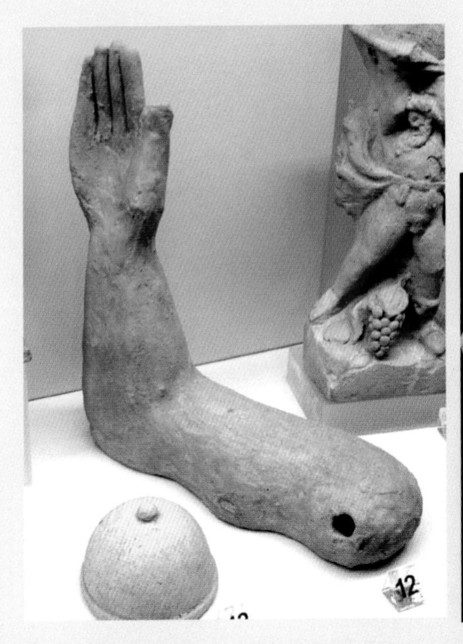

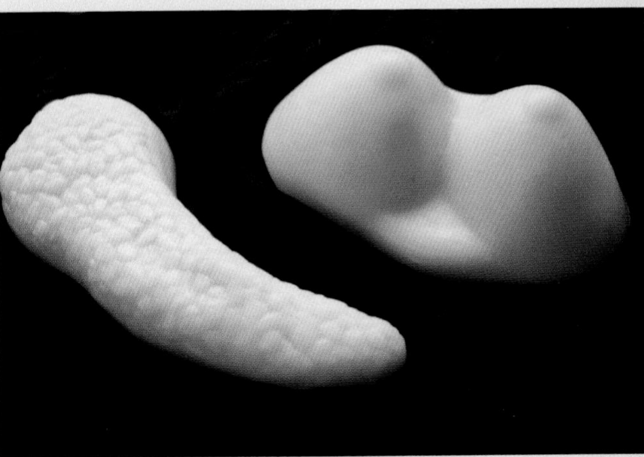

Church of the Nativity in Bethlehem, which is located atop a grotto long believed to be the birthplace of Jesus. Not far from there is a Milk Grotto (Magharet Sitti Mariam, or the "Grotto of the Lady Mary"), where according to tradition, the Holy Family took refuge during the "Slaughter of the Innocents" before their flight into Egypt. Legend has it that as Mary was nursing Jesus, a drop of her breast milk fell to the ground, turning the cave floor white. Today, Christian and Muslim women come to the cave to buy packets of powdered stone collected from the chamber, which when mixed with water is thought to enhance the quality of their breast milk. Stones from the cave, when placed under a mattress, are believed to enhance fertility.

VOTIVE OFFERINGS: THEN AND NOW
LEFT: Etruscan votive offerings from the IV century B.C. made in terracotta are exactly like the ones made of wax (RIGHT) which can be bought today. Body part votives were offered to indicate the type of malady to be healed or to give thanks for a recovery received.

In Italy, many of the ancient pagan milk caves have witnessed renewed use in the centuries following the arrival of Christianity with more than a dozen such grottoes serving as sanctuaries of the Archangel St. Michael, a valiant warrior who triumphed over a dragon-like Satan, banishing him to Hell. The Emperor Constantine (r. 306-337 A.D.) established the first sanctuary to the archangel in the

Near East, the Michaelion, which was built atop a pagan temple just north of Constantinople. The saint is said to have appeared in several caves, most notably in 490 A.D. within a grotto at Monte Sant'Angelo in the Apulian province of Foggia. Being identified as both a holy warrior and as a healer, St. Michael came to embody all of the attributes associated with Italy's earliest grotto cult sites. And not surprisingly, drip waters collected in fonts and stones collected from the sacred cavern continue to be sought for their miraculous powers. In his book, *Old Calabria* (1915), Norman Douglas quotes a 1910 pamphlet for a "Novena in Honor of St. Michael the Archangel" found at the site:

"It is very salutary to hold in esteem the stones, which are taken from the sacred cavern, partly because from immemorial times they have always been held in veneration by the faithful and also because they have been placed as relics of sepulchers and altars. Furthermore, it is known that during the plague which afflicted the kingdom of Naples in the year 1656 Monsignor G.A. Puccini, archbishop of Manfredonia, recommended everyone to carry devoutly on his person a fragment of the sacred stone whereby the majority were saved from the pestilence, and this augmented the devotion bestowed on them."

ST. MICHAEL'S CAVERN
This grotto has been used for religious rites since prehistoric times. It is now the sacred space dedicated to St. Michael. The sanctuary is deep within a limestone cavern and the walls still weep with water from the earth. RIGHT: The ceremonial blessing of water. Both the water and stones from the grotto are considered miraculous.

ST. FRANCIS' STIGMATA

The saint is portrayed in a rocky landscape by Giotto, who included natural elements in his works in keeping with the Franciscan notion that mountains were part of the sacred realm. The cleft in the rock is symbolic not only of the saint's simple life but of the wounds in his hands.

A New View Of Nature

The idea of a cavern as a dark and mysterious place suitable for religious rituals continued from antiquity to the Middle Ages, a time characterized by fear of nature and satanic forces. On the vanguard of a change in perspective toward the natural world was St. Francis of Assisi who is now associated with the environmental movement. He set forth the revolutionary philosophy that the Earth and all living creatures should be respected as creations of the Almighty. He lived in Umbria, a region which is still today, green, fertile and infused with a palpable spirituality. He constructed a series of monasteries which were one day's walk away from each other, situated in forests or snuggled up against the sides of mountains. His own cell and bed were carved out of rock. The grotto of the monastery at La Verna was the place at which he received the stigmata of Christ in 1224. An anthology of stories called the *Fioretti di San Francesco* (The Little Flowers of St. Francis) compiled a century later, describes the event:

> *"considering the form of the mountain and marveling at the exceeding great clefts and caverns in the mighty rocks, he betook himself to prayer and it was revealed to him that those clefts... had been miraculously made at the hour of the Passion of Christ when, according to the gospel, the rocks were rent asunder."*

St. Francis' affinity for Earthly elements influenced many of the artists of the time such as Giotto, who included rocky landscapes, mountains and forests in his frescoes as a way of presenting the newly popular idea of natural philosophy. The Franciscans continued to promote mountains as being vital in the sacred ritual, promulgating the idea that they would provide a nearness to God and a source of divine inspiration. Indeed, the clefts in the mountain were evocative of the wounds in St. Francis' hands and caused a wave of the faithful to make a pilgrimage to the grotto at La Verna to view the sacred recesses.

The inclusion of natural elements was not limited to visual art but to literature as well. Dante, in the *Divine Comedy* sets the scene of the entrance to Hell in a dark, foreboding, funnel-shapped cavern which is divided into ten concentric circles. He calls the eighth, the *Malebolge*, which means "evil ditch." Causeway type bridges extend from the outer circumference, much like the spokes of a wheel and meet in the center of the deepest, final circle of Hell. Dante places sinners in various levels of the cavernous inferno, with the worst of them relegated to the lowest.

Despite the revolutionary efforts made by St. Francis, until the Renaissance, nature was still considered mysterious—its ways unfathomable. One of the problems was that the texts available

FAKE GROTTO: THE ULTIMATE KNOCK-OFF
The wealthy of the Renaissance wanted a grotto that appeared real. They dug out the earth, created mountains and hired renowned architects and artists to aid in their construction. They spent enormous sums of money to import exotic rocks, shells and plants so they would be decorated authentically—the more realistic, the more prestigious.

for study were for the most part, ancient or ecclesiastical. Aristotle, Pliny and others formed the basis of natural thought and their ideas had not been altered or challenged in 1,500 years. With the dawn of the Renaissance however, a shift in thinking resulted in massive changes in many areas, and nature was one of them. From the end of the 1400s to the beginning of the 1500s a new cultural climate arose in which philosophy was being reexamined—with Plato, rather than Aristotle, being the preferred philosopher.

It was in this "reborn" cultural climate that the quest for the new, the innovative, took hold and the idea of a grotto emerged in a new form in an unlikely place. It was Isabella d'Este, close friend of Leonardo da Vinci, who decided to build an alcove adjoining her study which she called a "grotto." While we might think of this as an isolated case of home redecoration, the fact that she and her husband ruled the Duchy of Mantua and were considered trend setters made the of building a grotto in their palazzo newsworthy and was cited in correspondence for years afterward. And in fact, the word "grotto" between the late 1400s and beginning of the 1500s did not mean an alcove in a garden, but rather a room located in a building.

The change in location of a grotto from inside to outside occurred due to the writings of two Renaissance personalities. Leon Battista Alberti was the first to write a treatise in 1450, *De Re Aedificatoria* (On the Art of Building), which gave detailed descriptions of ancient grottoes. He was familiar with their characteristics from accounts in Ovid's *Metamorphosis III*, 159 and *Fasti II*, 315. Alberti's suggestion for decorating the newly constructed grottoes with shells, coral and colorful glass beads was a result of his humanistic perspective and represented the cultural shift occurring in the Renaissance.

It was the Sienese engineer, Francesco di Giorgio Martini, who wrote the first treatise (*Trattato di Architettura civile e militare*) on the theory of a grotto in the mid XV century. He proposed a garden plan which called for a "grotto" and a chapel adjoining it. Federico da Montefeltro, the Duke of Urbino, followed Martini's advice and constructed one. Not to be outdone, style-conscious Isabella d'Este rushed to construct her own at Palazzo Te in Mantua.

In the second half of the 1400s, artists chose a new style which favored the insertion of rocks, caverns and grottoes in contemporary iconography. Even writers developed this theme. The humanists like Poggio Bracciolini described ancient relics being surrounded by rocks and vegetation. Grottoes became more popular in literature and painting before architecture itself. To understand this, we have to examine the Platonic and Neoplatonic thought which was in vogue. Plato, in Book VII of the *Republic* states "*behold men as if dwelling in a subterranean cavern.*" It seems like people wanted to take him literally.

IMITATING NATURE
This marvelously preserved grotto by Buontalenti in Florence's Boboli Gardens is characteristic of the care and artistry demonstrated by craftsmen of the time. A copy of a Michelangelo sculpture, to the left, the Bearded Prisoner, emerges from imitation stalactites.

Of Things Grotesque

In the late 1400s a young man fell down a crevice on a Roman hill. Amazingly, he found himself in a grotto adorned with ornate frescoes. As word spread about their beauty, artists who were working at the Vatican decided to have a look. Slowly making their way down, clinging to a knotted rope were Michelangelo, Raphael and Pinturicchio. They were astounded to see the paintings of classical antiquity. What was soon realized was that this was not a grotto, but a room in Nero's villa, Domus Aurea, which having been buried under 1,500 years of fill, took on the aspect of a grotto. Further exploration revealed room after room of the magnificent villa which was started after the great fire of Rome in 64 A.D. The walls of each room

TIBERIUS' DINING GROTTO

About 70 miles south of Rome, in Sperlonga, the Emperor Tiberius chose a special location to build his seaside villa in the I century A.D. The site had a natural limestone grotto, seen above, complete with dripping stalactites and adorned with enormous sculptures. He had a series of pools built, (foreground) with a dining platform in the middle (center, grass is growing from it). A number of bridges connected the dining platform to the grotto so guests and servants could come and go and performances within the grotto could be enjoyed while dining under the stars.

were decorated in the extravagant style of ancient Rome. Quick to realize a "new" form of decorative art, Raphael used it immediately in the loggias which are part of the series of Raphael's rooms in the Vatican. He called this new style "grotesque" from the Latin *grupta*, meaning crypt or grotto, in homage to the place in which he received his inspiration.

Real Or Imagined: The Grotto As The Ultimate Must-Have

To understand how grottoes moved from inside to outdoors, we have to refer to Platonic and Neoplatonic texts which amply discuss caverns. Often they are described as mysterious places associated with an epiphany. The fascination with grottoes came about as Renaissance scholars rediscovered Classical

SUMPTUOUS STALACTITES
An example of the beauty of the imitation grottoes which have survived. They are masterpieces of artistry using a combination of real objects such as shells, glass and stone and artificial materials such as the stalactite formations made of various mixtures of stucco and cement.

texts and wished to emulate the ancients. The most influential of these was *Cavern of the Nymphs* by Porphyry (233-309 A.D.), a philosopher and one of the founders of Neoplatonism born in the Phoenician port of Tyre. His treatise became the bible for those wishing to construct the perfect grotto, physically and metaphorically that is, as this was the place where one could be transformed. Porphyry described the grotto much like the cosmos and defined it as warm, dark and moist. The theme was not new, having been treated in Plato's *Timaeus* (360 B.C.), but the timing of the republication of Porphyry's work in Rome in 1518 was perfect and it set off a wave of creativity.

Porphyry's text was rich with examples and served as a model for the grottoes and gardens of the 1500s. He describes the nymphs who inhabit these primordial places as symbols of souls who thrive on humidity. A naiad was a type of nymph who presided over springs, streams, brooks, wells and fountains; all of which had sacred connotations since the dawn of time. The essence of a naiad was bound to her spring, so if her source of water dried up, she would die.

The diffusion of Platonic ideas among the wealthy morphed into an unnatural state, becoming superficial. At the start of the 1400s a new appreciation of nature, with all its attendant mystical implications began. But within a hundred years, the spiritual element was lost amid a frenzy to attain enlightenment via materialism. The rich and powerful of the Renaissance went crazy for grottoes. They built fake ones, but insisted on them looking natural to simulate "cosmic" material. This obsession to build was based on dual objectives: emulate the esteemed ancients and flaunt their wealth. No expense was spared in creating some of the most extravagant garden grottoes—for which stalactites from real caves and seashells, glass and stone were imported to adorn the recesses while sculptures—often of mythological subjects—were commissioned

FEMALE FERTILITY **Reminiscent of prehistoric fertility sites where offerings were made for conception and abundant breastmilk, the image of the Fountain of Diana of Ephesus, located in Villa d'Este at Tivoli, is still strikingly evocative.**

from leading artists. Lavish parties were held, and these gardens became an important meeting place for artists, intellectuals and politicians. Word of their beauty spread and visitors from all over Europe came to admire and be entertained. They would serve as the model for future generations of gardens in France and England.

Under Francesco I de' Medici (1541-1587) of Florence, the most famous grottoes of the Renaissance were built: the Boboli Gardens grotto and the one in neighboring Pratolino. These harken back to Porphyry's theme of the needs of the soul, yet are filled with evidence of conspicuous consumption in the extravagant man-made objects and mechanical innovations. The Boboli grotto characterizes man's journey, being symbolically metamorphosed from the Earthly dimension to the divine. At Pratolino, access to a subterranean cavern brings one to a place thought to have been inhabited by souls who lived among the rocks. The placement of a labyrinth nearby is a Porphyrian idea symbolizing the start of a journey of spiritual initiation. Comfortingly implicit for those embarking on this path of transformation was that the route could be immediately reversed.

Ischia: An Ancient Island Apothecary

It has been 14 years since I began studying the waters of Ischia for the good of the world. I study each place and examine the minerals and the caverns and finally, with great care, I observe the stupendous effects of these waters on sick bodies and the beneficial effects on those with sane bodies. I have withstood difficult voyages by ship to reach the island, and feared the pirates, which frequented these waters, yet I continued to come, of my own volition to study the divinely beneficial effects of these waters on those who need them most.

—Giulio Iasolino *De' rimedi naturali che sono nell'isola Pithaecusa, oggi detta Ischia ("Natural remedies on the island of Pithaecusa, known today as Ischia")* (1588)

For more than two millennia, the thermal waters of Ischia—a 46-square-kilometer volcanic island at the northern entrance of the Gulf of Naples—have been known for their extraordinary healing properties. Here, waters issuing from more than a hundred springs are able to ameliorate ailments such as rheumatism, arthritis, osteoporosis, sciatica and a host of respiratory illnesses, allergies and skin conditions.

Ischia's waters' magical properties were first recognized by the ancient Greeks, who established a trading hub on the island in the VIII century B.C. They called it Pithekousai (πιθηκοσσι), the name likely a reference to the island's high-quality clays, which were prized for the manufacture of pithoi (πιθοι), the large storage jars used for transporting wine, oil, honey and garum (fish sauce) throughout the Mediterranean world. Others say the name derives from Greek myth, where

In 1759 farmers near Barano discovered a dozen marble votive reliefs dedicated to the Nymphs of Nitrodi and dated from the I century B.C. to the III century A.D. We can see the names of those who had been cured and made these offerings, such as Poppaea Augusta, who thanked Apollo and the Nymphs. In the I century A.D. a doctor Menippus, "Menippos iatròs upalpinos" (Menippos, from the near side of the Alps) came to Nitrodi from Northern Italy. Later, two other doctors; "Aur (elius) Monnus" and "Num (erius) Fabius" came to study the waters and treat patients.

a race of mischievous forest creatures called *Cercopes* were turned into monkeys (pithecanthropus) by Zeus and banished to volcanic areas, one of which was Ischia.

It seems like a veritable who's who of Greek and Roman poets and commentators such as Virgil, Homer, Ovid, Statius and Lucan, to name a few, described Ischia as having been inhabited by giants. One in particular, named Typhon was the monstrous son of Gaea, goddess of the Earth and Tartarus, god of the Underworld. He was the personification of volcanism since flames gushed from his mouth. In a presumptuous move, he challenged Zeus to a fight, which he lost. Seeking a suitable punishment for the hothead, Zeus buried him under the island of Ischia. Unable to endure the humiliation,

The Trifecta of Thermal Therapy

A plethora of ailments could be remedied by the right combination of these components. The results have been well documented, not only from a medical perspective, but from a beauty standpoint, leading these waters to be known as "the fountain of youth."

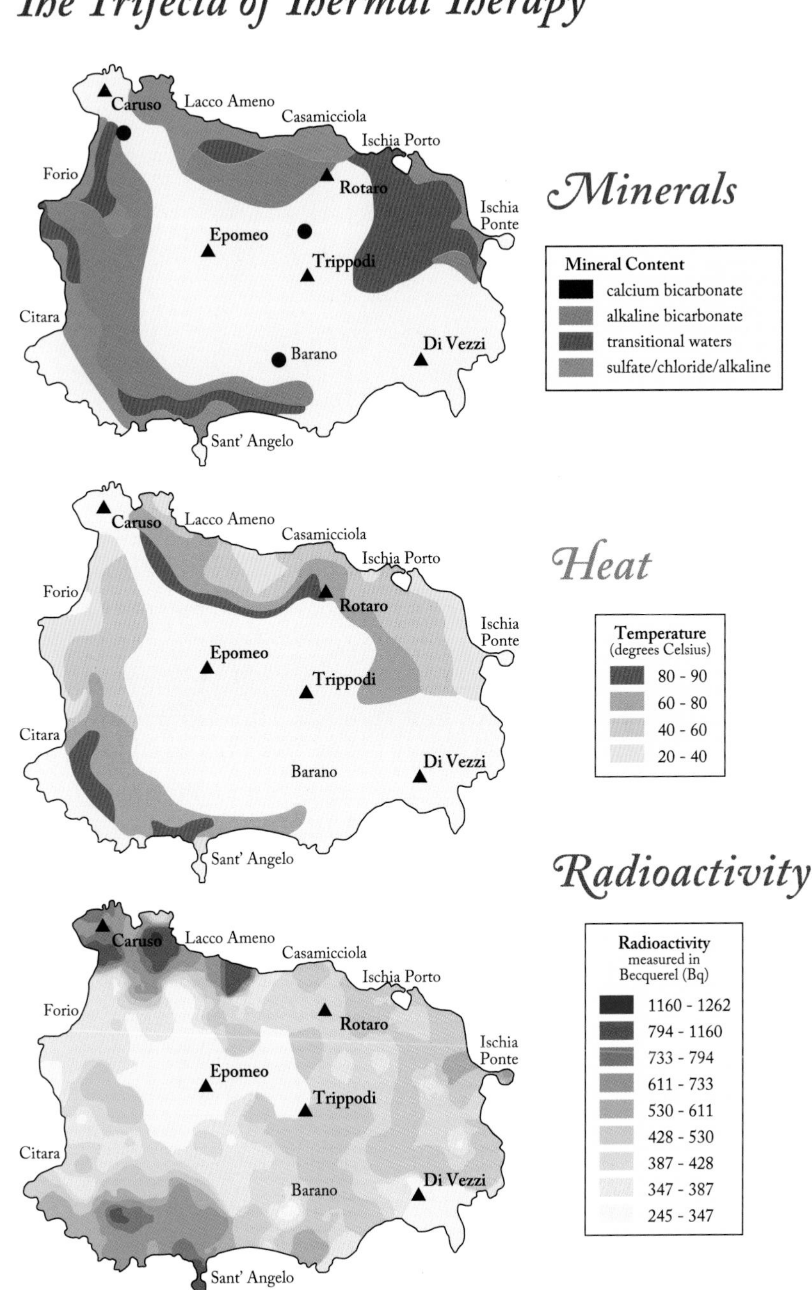

A votive offering from Nitrodi in the form of a triptych. In the lateral spaces, the Dioscuri. At the center, the three nymphs depicted as the graces: charm, beauty and creativity. At the bottom, the personification of a spring, which was considered sacred. The ancients believed that a healthy body was better able to spiritually connect with the gods.

the angry giant struggled to free himself, spewing forth flames and boiling water and shaking the earth with his restless movement. Frustrated that he could not escape his fate, he began to cry so intensely that Aphrodite was moved to free him, turning his tears into waters with healing powers.

Not surprisingly, temples and shrines were erected around the springs dedicated not only to Aphrodite but to Apollo, a god with deep associations with health and healing. Apollo was known by a number of epithets, including Akesios, from the Greek word for "healing," which may have in part inspired the island's current name. Ischia's healing waters attracted people seeking wellness, however, the island's penchant for sudden seismic outbursts discouraged ambitious building programs and continuous settlement.

Although long attributed to divine machinations, Ischia's pharmacological bounty is a byproduct of the island's violent birth and ongoing seismic activity. Formed over a 150,000-year-period, this complex volcanic system began with an eruption that left a massive caldera, the perimeter of which roughly encircles the island. In time, eruption after eruption left its mark on the island in the form of craters and lava flows. Some 39,000 years ago, a forceful volcanic eruption centered at Pozzuoli on the mainland caused all but the southeastern coast of Ischia to collapse into the sea. In time, however, the inflation of the magma chamber offshore resulted in the dramatic uplift of a dome of marine sediments lying on the seafloor—the Green Tuff ignimbrite and clays that make up much of the island interior we see today. Throughout the Holocene (ca. 12,000 years ago to present) smaller volcanoes erupted around the

The Nymphs of Nitrodi hold shells and vessels to distribute the healing water of the spring. The father of medicine, Hippocrates, praised the properties of mineral water and hot springs to heal the body and render it more youthful. Pliny and Strabo echoed the Greeks' appreciation for the wondrous waters of Ischia.

dome, with the last major volcanic event on Ischia taking place in 1302 A.D. Earthquakes, however, continued to rock the island until the late XIX century.

Geologically speaking, the 789-meter-high Mt. Epomeo, which covers some 35 percent of the Ischia, is one of several side by side "horsts" or wedges of a once-contiguous seafloor that was uplifted then split by tectonic activity. The horsts are separated by faults through which flow gases, vapors and the thermal waters for which the island is famous.

Emanating from underground reservoirs fed by rainwater, the waters are warmed by heat sources located deep within the Earth. The waters are transformed into steam, which rises to the surface, enriched along its journey by the minerals contained in the soil. As a result, the waters gushing out are alkaline or acidic and contain varying amounts of calcium, magnesium, hydrogen carbonate, sodium, sulfur, iodine, chlorine, iron, potassium and microelements of other active substances, including radon. It is the disparate chemical composition of the overlying sediments through which the steam and thermal waters make their way that dictates their healing properties.

In his two-volume 1588 treatise, Calabrian physician Giulio Iasolino provided the first analysis of the island's spring waters, classifying each font and spa according to its composition and distinctive properties. Iasolino found 35 natural springs capable of curing arthritis, gout, sterility in women and restoring virility to old men, but out of all of the waters he "happily experimented with" it was the Gurgitello Spring, in Casamicciola, which he like the best, describing it's attributes as follows:

"Now let us discuss the most precious water

IN-DEPTH **Ischia: An Ancient Island Apothecary**

The island of Ischia lounges languidly in the Gulf of Naples, yet it still emits thermal waters, vapors and mud each and every day.

source, commonly referred to as Gurgitello, and the empirical demonstration of the efficacy of its waters on many illnesses. It is reputed to make sterile women able to conceive, make old men virile again, soothe stomach irritations, aid in passing kidney stones, heal the liver, cure scabies, restore one's appetite and as the local inhabitants say, aided in healing an open wound of someone who had a metal splinter in his chest for more than one year."

Ischia's water is rich in noble elements and mineral salts: sodium, one of the basic elements for the vital activity of living beings, potassium, essential for muscle dynamics, sulfur, essential for joint elasticity and calcium, which has a sedative action on the nervous system. The radioactive nature of the waters was discovered in 1918, when Marie Curie came to Ischia with fellow scientists to study the thermal springs. She determined that the waters bore various components of radium, radon, thorium, uranium and actinium. The most important therapeutic element is radon, a gas dissolved in the water originating from an alpha particle emanated by an atom of radium. The radioactivity is so low that it is not harmful, a sheet of paper is enough to stop it and it has a short half-life so it isn't able to bio-accumulate.

Even now, Iasolino's publication remains the consummate treatise on medical hydrology. And while this early study focused on the health effects of the springs, Iasolino later noted the aesthetic benefits as well, writing that, *"Women who regularly wash their clothes in it and use it for other purposes benefit. Such women are beautiful and keep their bodies healthy."*

Today, thanks to the restless Earth, one can still enjoy a wellness experience 150,000 years in the making.

I roamed the countryside searching for answers to things I did not understand.... These questions and other strange phenomena engaged my thoughts throughout my life.

Leonardo da Vinci

And so, follow

Leonardo da Vinci

and ask…

What secrets will the Earth

reveal to me today?

Bibliography

INTRODUCTION.
ITALY: A LAND BORN OF THE SEA *page 1*

Baldi, Antonio, Napoli Geologica, Napoli, Tempo Lungo Editore, 1998.

Angelelli, Francesco and Rossi, Roberta, Catalogue of Types preserved in Paleontological Collections of APAT, Rome, 2004.

Crescenti, V., D'Offizi, S., Merlino, S., Sacchi, L., Geology of Italy, Special Volume of the Italian Geological Society for the IGC 32, 2004.

Istituto Nazionale di Geofisica e Vulcanologia, Italian geophysical and volcanic website: www.ingv.it/en/

Minelli, Giorgio and Foglietta, Mauro, "The geology of Italy: a brief overview," in Periodico di Mineralogia,72, 2003.

Pizzorusso, Ann C., The Ground We Walk On, Earth Scape, www.napoliunplugged.com.

Spalluto, Luigi and Caffau, Mauro, "Stratigraphy of the mid-Cretaceous shallow-water limestones of the Apulia Carbonate Platform," Italian Journal of Geoscience vol. 129, No. 3, 2010.

I. A LANDSCAPE DIVINE:
ETRUSCANS AND THEIR ENVIRONMENT *page 11*

Becker, Robert O. and Selden, Gary, The Body Electric, New York, William Morrow, 1985.

Benvenuti, M., Boni, M. and Meinert, L., "Skarn Deposits in Southern Tuscany and Elba Island Central Italy" in Proceedings of the 32nd International Geological Congress, Volume 2. Rome: APAT (Italian Agency for the Environmental Protection and Technical Services), 2004.

Berg, H., "A Short Account of 'LightningTubes,'" Scientific American, August 15, 1908.

Bonfante, Giuliano and Bonfante, Larissa, The Etruscan Language, An Introduction (revised edition). Manchester, Manchester University Press, 2002.

Bonfante, Larissa and Swadling, Judith, Etruscan Myths. Austin, University of Texas Press/British Museum Press, 2006.

Chiappini, Massimo, "Shaded relief magnetic anomaly map of Italy and surrounding marine areas," Annali di Geofisica, Vol. 43, No 5, October 2000.

De Grummond, Nancy Thomson and Edlund-Berry, I. (Editors), The Archaeology of Sanctuaries and Ritual in Etruria, Journal of Roman Archaeology, Supplement 81. Portsmouth, RI: JRA, 2011.

De Grummond, Nancy Thomson and Simon, Erika (Editors), The Religion of the Etruscans. Austin, University of Texas Press, 2006.

Derr, J. S. and Persinger, M. A., "Luminous phenomena and earthquakes in southern Washington," Experientia, 42, pp.991–999, 1986.

Domenici, Viviano, Archeologia e Medicina: Gli Etruschi e il Fegato di Piacenza, Piacenza: Camillo Corvi, SPA, 1986.

Frank, Tenney, An Economic History of Rome (Second Edition Revised; orginally published in 1927), Kitchener, ON, Batoche Books Limited, 2004.

Force, E. R. and McFadgen, B. G., "Tectonic environments of ancient civilizations: opportunities for archaeoseismological and anthropological studies" in Ancient Earthquakes, M. Sintubin, I. S. Stewart, T. M. Niemi, and E. Altunel, eds.: Geological Society of America Special Paper 471, 2010.

Freund, Friedemann T., "Rocks That Crackle and Sparkle and Glow: Strange Pre-Earthquake Phenomena," Journal of Scientific Exploration, Vol. 17, No. 1, pp. 37–71, 2003.

Ghirotto, S., Tassi, F., Fumagalli, E., Colonna V., Sandionigi, A., et al. "Origins and Evolution of the Etruscans' mtDNA" Plos One 8(2):e55519 (2013).

Grigor'ev, A. I., Grigor'eva, I. D., and Shiryaeva, S. O., "Ball lightning penetration into closed rooms: 43 eye witness accounts." Journal of Scientific Exploration, 6, pp.261–279. 1992.

Harrison, Adrian P. and Turfa, Jean M., "Were natural forms of treatment for Fasciola hepatica available to the Etruscans?" in International Journal of Medical Sciences 7 (6). New South Wales, Australia: Ivyspring International Publisher, 2010.

Istituto Nazionale di Geofisica e Vulcanologia, Italian geophysical and volcanic website: www.ingv.it/en/

Pellecchia, Marco, et al., "The mystery of Etruscan origins: novel clues from Bos taurus mitochondrial DNA," in Proceedings of the Royal Society B. London: Royal Society, 2007.

Pliny the Elder, The Natural History, English translation by John Bostock, M.D. London: Taylor and Francis, 1855. Courtesy Perseus Project.

Stenhoff, M., Ball Lightning: An Unsolved Problem in Atmospheric Physics. New York: Kluwer Academic/Plenum Publishers, 1999.

Titus Livius (Livy), Ab urbe condita libri (History of Rome). English Translation by Rev. Canon Roberts. New York: E.P. Dutton and Co., 1912. Courtesy Perseus Project.

II. HELL ON EARTH: INTO VIRGIL'S UNDERWORLD *page 59*

Campi Flegrei: mito, storia, realtà, Catalogue of exhibit, Napoli, Electa Editore, 2006.

Comparetti, Domenico, Vergil in the Middle Ages, Princeton University Press, 1997.

Desimone, R., Il segno di Virgilio, Sezione Editoriale Puteoli, 1982.

Fairclough, H.R., Virgil Aeneid I-VI, Harvard University Press, 1999.

Fiore, Giuseppe, La Grotta del Cane alla Conca di Agnano, Napoli, RCEmultimedia, 2012.

Hobel, Sigfrido, Napoli vista dal mare, Ascultur Campania, 2001.

Infusino, Giampaolo, Storia, Miti e Leggende dei Campi Flegrei, Lito-Rama, 1995.

Mauri, Amedeo, I Campi Flegrei: dal sepolcro di Virgilio all'antro di Cuma, Roma, La Libreria dello stato, 1934.

Pizzorusso, Ann C., The Burning Fields, Earth Scape, www.napoliunplugged.com.

Rosi, M., Sbrana, A. and Principe, C., "The Phlegraean Fields; Structural evolution, volcanic history and eruptive mechanisms," in M.F. Sheridan and F. Barberi (eds.), Explosive Volcanism, Journal of Volcanology and Geothermal Research, vol. 17, Elsevier Science, Ltd., 1983.

Sirpettino, M., Mito e mistero nei Campi Flegrei, Franco di Mauro Editore, 1991.

Spargo, John Webster, Virgil the Necromancer: Studies in Virgilian Legends, Harvard University Press, 1934.

CHRIST STOPPED AT LAKE AVERNO *page 78*

Baldi, Antonio, Napoli geologica, Napoli, Tempo Lungo Editore, 1998.

Campi Flegrei: mito, storia, realtà, Catalogue of exhibit, Napoli, Electa Editore, 2006.

D'Eboli, Pietro, De balneis Terrae Laboris, codex in the Biblioteca Angelica of Rome.

PYROCLASTIC POETS *page 80*

Benjamin, Sandra, Sicily: Three Thousand Years of Human History, Hanover, N.H., Steerforth Press, 2006.

Dickinson, Emily, The Complete Poems of Emily Dickinson, Wetzsteon, Rachel, ed. Barnes and Noble, 2003.

Istituto Nazionale di Geofisica e Vulcanologia, Italian geophysical and volcanic website: www.ingv.it/en/

Pizzorusso, Ann C., The Umbilical Cord, Earth Scape, www.napoliunplugged.com.

Reiss, Tom, The Black Count: Glory, Revolution, Betrayal, and the Real Count of Monte Cristo, Crown, 2013.

Shelley, Percy Bysshe, Complete Poems, Modern Library, 1994.

Shelley, Percy Bysshe, Shelley's Poetry and Prose, Fraistat, Neil and Reiman, Donald, ed., Norton, 2002.

Tomalin, Claire, Charles Dickens: A Life, New York, Penguin, 2012.

Twain, Mark, The Innocents Abroad, reprint 2012.

Virgil, The Georgics, Wilkinson, L.P., ed., Penguin, 1982.

Von Goethe, Johan Wolfgang, Auden, W.H. and Mayer, Elizabeth, trans., Italian Journey: 1796-1788, San Francisco, North Point Press, 1982.

Von Goethe, Johan Wolfgang, The Collected Works of Johan Wolfgang von Goethe, Bowring, Edgar Alfred, et. al., trans., Halcyon Classics, 2010.

Wollstonecraft, Mary Shelley, Frankenstein, reprint 2011.

Wollstonecraft, Mary Shelley, Original Stories, Lucas, Edward Verrall, ed. reprint 2010.

RISE AND FALL OF A ROMAN RESORT *page 85*

Baldi, Antonio, Napoli geologica, Napoli, Tempo Lungo Editore, 1998.

Campi Flegrei: mito, storia, realtà, Catalogue of exhibit, Napoli, Electa Editore, 2006.

Istituto Nazionale di Geofisica e Vulcanologia, Italian geophysical and volcanic website: www.ingv.it/en/

Mancini, Enzo, Flegree, Isole dei Verdi Vulcani, Napoli, Mursia, 1980.

Mauri, Amedeo, I Campi Flegrei: dal sepolcro di Virgilio all'antro di Cuma, Roma, La Libreria dello stato, 1934.

Museo Archeologico di Napoli, Greco,Gabriella, ed. Il Sole 24 Ore, Napoli, Electa, 2005.

Pane, Roberto, Virgilio e i Campi Flegrei, Napoli, Adriano Gallina Editore, 1981.

III. PARADISE BEJEWLED *page 91*

Bari, Hubert, Cardona, Caterina, Parodi, Gian Carlo, Diamanti, arte, storia, scienza, De Luca Editore d'Arte, 2002.

Budge, E.A. Wallis, Amulets and Superstitions, Dover, 2011.

Caley, Earle R. and Richards, John F. C., Theophrastus' On Stones: Introduction, Greek Text, English Translation, and Commentary. Ohio State University, 1956.

Freccero, John, Dante: The Poetics of Conversion, Harvard University Press, 1988.

Kunz, George Frederick, Magic of Jewels and Charms, Kessinger reprint, 2003.

Kunz, George Frederick, The Curious Lore of Precious Stones, Dover, 1971.

Mandelbaum, Allen, The Divine Comedy of Dante Alighieri: Inferno, Bantam Books, 1980.

Mandelbaum, Allen, The Divine Comedy of Dante Alighieri: Purgatorio, Bantam Books, 1982.

Mandelbaum, Allen, The Divine Comedy of Dante Alighieri: Paradiso, Bantam Books, 1984.

Mottana, Annibale, "Italian gemology during the Renaissance: A step toward modern mineralogy," in The Origins of Geology in Italy, The Geological Society of America, Special Paper 411, 2006.

Riddle, John M., Marbode of Rennes' (1035-1123) De lapidibus: Considered as a medical treatise with text, commentary, and C.W. King's translation, together with text and translation of Marbode's minor works on stones. Sudhoffs Archiv, 1977.

IV. LEONARDO DA VINCI: ON THE NATURE OF THINGS *page 143*

Ackerman, Steven A. and Knox, John, Meteorology, 3rd. Ed. Sudbury, Ma., Jones & Bartlett Learning, 2011.

DaVinci, Leonardo, Trattato della Pittura, section on painting in Codex Urbinas, Vatican City, Vatican Library, compiled ca. 1542.

Kemp, Martin, (ed.), Leonardo on Painting, New Haven, Yale University Press, 1989.

Leonardo da Vinci: Codex Leicester—A Masterpiece of Science, New York: American Museum of Natural History, 1996.

Ravaisson-Mollien, C, (ed.), Les Manuscrits de Leonard de Vinci, Manuscrit A. (etc.) de l'Institut de France, (6 vols.) Paris, 1881-91.

Pedretti, Carlo, Leonardo da Vinci: Nature Studies from the Royal Library at Windsor Castle, Johnson Reprint, 1980.

MacCurdy, Edward, The Notebooks of Leonardo da Vinci, New York, Reynal & Hitchcock, 1938.

Reti, Ladislao, The Unknown Leonardo, New York, Abrams, 1990.

Richter, Jean-Paul (ed.), The Literary Works of Leonardo da Vinci, New York, Dover Editions, 1970.

LEONARDO'S POST-DILUVIAN WORLD *page 154*

Clark, Kenneth, Landscape into Art, New York, Harper Collins, 1991.

MacCurdy, Edward (ed.), The Notebooks of Leonardo da Vinci, New York, Reynal & Hitchcock, 1938.

THE EARTH'S PAINT BOX *page 158*

Beck, James, Italian Renaissance Painting, Konemann UK Ltd. 2nd revised ed. 1999.

Burchard, Ernest F. The Production of Mineral Paints 1907, Kessinger Legacy Reprints, 2010.

Dunstan, Bernard, Painting Methods of the Impressionists, Watson-Guptill Publications Inc. US, 2nd revised ed. 1983.

McGraw-Hill, Dictionary of Geology and Mineralogy, 2nd ed. 2003.

V. LEONARDO'S GEOLOGY: A TALE OF TWO PAINTINGS *page 163*

Brizio, Anna Maria, Scritti Scelti di Leonardo da Vinci, Torino: Unione Tipografico, 1968.

Della Chiesa, Angela Ottino, The Complete Paintings of Leonardo da Vinci, New York, Penguin, 1967.

Hope, Charles, "The Wrong Leonardo?" The New York Review of Books, February 9, 2012.

MacCurdy, Edward (ed.), The Notebooks of Leonardo da Vinci, New York, Reynal & Hitchcock, 1938.

Nardini, Bruno, Vita di Leonardo, Firenze: Giunti Marzocco, 1978.

Pater, Walter Horatio, The Renaissance: Studies in Art and Poetry, Teddington, UK, The Echo Library 2006.

Pedretti, Carlo, Leonardo, A study in Chronology and Style, Los Angeles, University of California Press, 1973.

Pedretti, Carlo, Leonardo Da Vinci: Nature Studies from the Royal Library at Windsor Castle, Washington: University of Washington Press, 1983.

Wasserman, Jack, Leonardo da Vinci, New York, Abrams, 1984.

Pizzorusso, Ann C. "A geologic comparison of the two versions of the Virgin of the Rocks," www.leonardosgeology.com

Pizzorusso, Ann C., "Leonardo's Geology: The Authenticity of the Virgin of the Rocks," Leonardo Journal, Vol. 29. No. 3, The MIT Press, 1996.

Pizzorusso, Ann, "Leonardo's geology: a key of identifying the works of Boltraffio, d'Oggiono, and other artists," Raccolta Vinciana, Vol. 27.1997.

Pizzorusso, Ann C., "Could the Louvre's Virgin and St. Anne provide the proof the (London) National Gallery's Virgin of the Rocks is not by Leonardo da Vinci?" http://artwatchuk.wordpress.com/2012/06/12/12-june-2012/

VI. RECESSES OF THE MIND AND SOUL *page 180*

Alberti, Leon Battista, De Re Aedificatoria, On the Art of Building in Ten Books, Cambridge, MIT Press, 1988.

Arzarello, Marta and Peretto, Carlo, "Out of Africa: The first evidence of Italian peninsula occupation," Quarternary International (2010).

Arrighi, Simona and Borgia, Valentina, "Analisi funzionale degli strumenti litici di corredo alle sepilture II e III di Grotta Paglicci (Rignano Garganico—Foggia)," Arte e Spitirualità, Annali dell'Università degli Studi di Ferrara, Museologia Scientifica e Naturalistica, (Volume speciale, 2007).

Bianco, Salvatore, "Il culto delle acque nella Preistoria," Archeologia dell'Acqua in Basilicata, Potenza, 1999.

Castelli, Patrizia, "L'antro delle Ninfe" in Artifici d'Acque e Giardini, La Cultura delle Grotte e dei ninfei in Italia e in Europa, Atti del V Convegno Internazionale sui Parchi e Giardini Storici, Firenze, 1999.

Douglas, Norman, Old Calabria (1915), Marlboro Press, 1996.

Elster, Ernestine S., "Grotta Scaloria in Genoa with an International Cast," Backdirt, Cotsen Institute, UCLA, 2009.

Elster, Ernestine S.; Isetti, Eugenia; and Traverso, Antonella, "Nuove evidenze di studio dal sito de Grotta Scaloria (Fg)," Atti del 28° Convegno Nazionale sulla Preistoria-Protostoria-Storia della Daunia (San Severo, 2008)

Ferriolo, Massimo Venturi, "La grotta, le ninfe e il paesaggio della Grande Dea" in Artifici d'Acque e Giardini, La Cultura delle Grotte e dei ninfei in Italia e in Europa, Atti del V Convegno Internazionale sui Parchi e Giardini Storici, Firenze, 1999.

Galletti, Giorgio, "La genesi della Grotta Grande di Boboli" in Artifici d'Acque e Giardini, La Cultura delle Grotte e dei ninfei in Italia e in Europa, Atti del V Convegno Internazionale sui Parchi e Giardini Storici, Firenze, 1999.

Guglielmo, Enrico, "Il Ninfeo romano imperial di Punta Epitaffio a Baia ed il suo allestimento nel Museo Archeologico dei Campi Flegrei" in Artifici d'Acque e Giardini, La Cultura delle Grotte e dei ninfei in Italia e in Europa, Atti del V Convegno Internazionale sui Parchi e Giardini Storici, Firenze, 1999.

Homer, The Odyssey, translated by Richmond Lattimore, New York, Harper and Row, 1967.

Pietrogrande, Antonella, "Una spelonca di dolci acque amena" Grotte e ninfeo tra Umanesimo e Manierismo, in Artifici d'Acque e Giardini, La Cultura delle Grotte e dei ninfei in Italia e in Europa, Atti del V Convegno Internazionale sui Parchi e Giardini Storici, Firenze, 1999.

Porphyry of Tyre, Cavern of the Nymphs, translated by Thomas Taylor, Grand Rapids, Mi., Phanes Press, 1991.

Richter, Irma, A., (ed.) The Notebooks of Leonardo da Vinci, Oxford, UK, Oxford University Press, 1980.

St. Francis, The Little Flowers of St. Francis, translated by Thomas Okey, New York, Dover, 2003.

Toledano, Ralph, Francesco di Giorgio Martini, Pittore e Scultore, Milano, Mondadori, 1987.

Tiné, Santo, "Un culto neolitico delle acque nella Grotta Scaloria," Actes du Symposium International sur les religions de la préhistorie, Capo di Ponte, 1975.

Tiné, Santo; Isetti, Eugenia, "Culto Neolitico delle acque e recenti scavi nella grotta Scaloria," Bollettino di Palentologia Italiana, 1980.

Città di Manfredonia—Provincia di Foggia, www.commune.manfredonia.fg.it.aast/grottascaloria.htm (June 25, 2012).

Virgil, The Eclogues, translated by Guy Lee, London, Penguin Classics, 1980.

Whitehouse, Ruth, Underground Religion: Cult and Culture in Prehistoric Italy, University of London, 1992.

ISCHIA: AN ANCIENT ISLAND APOTHECARY *page 205*

Avino, R., et. al., Geochemical Investigations in soils and waters of Ischia Island, The Smithsonian/NASA Astrophysics Data System, EGS General Assembly, Nice, April, 2002.

Bartoli, G., et. al, Evaluation of the exposure levels to radioactivity in the hot-spring environment of the Island of Ischia during a year. U.S. National Library of Medicine, National Institutes of Health, 1989.

Baldi, Antonio, Napoli geologica, Napoli, Tempo Lungo Editore, 1998.

Bucher, Paul, Giulio Iasolino, Lacco Ameno, Ischia, Imagaenaria, 2000.

Campi Flegrei: mito, storia, realtà, Catalogue of exhibit, Napoli, Electa Editore, 2006.

Douglas, Norman, Isole d'estate: Ischia e Ponza, Balestriere, Giorgio, trans., Lacco Ameno, Ischia, Imagaenaria Editore, 2004.

Greco,Gabriella,(ed.), Museo Archeologico di Napoli, Il Sole 24 Ore, Napoli, Electa, 2005.

Iasolino, Giulio, De' remedi naturali che sono nell'isola di Pithecusa hoggi detta Ischia, reprint Lacco Ameno, Ischia, Imagaenaria, 2000.

Mancini, Enzo, Flegree Isole dei Verdi Vulcani, Napoli, Mursia, 1980.

Orioli, Giuseppe, Giro indipendente dell'isola d'Ischia, Lacco Ameno, Ischia, Imagaenaria, 2004.

Pizzorusso, Ann C. Ischia-Growing Up is hard to do, Earth Scape Naples, www.napoliunplugged.com, 2012.

Pizzorusso, Ann C. Ischia-Geologic Wonderland, Earth Scape Naples, www.napoliunplugged.com, 2012.

Pizzorusso, Ann C. Ischia-Mystery of the Thermal Waters, Earth Scape Naples, www.napoliunplugged.com, 2012.

Pizzorusso, Ann C. Ischia-Fountain of Youth, Earth Scape Naples, www.napoliunplugged.com, 2012.

Pugliese, M., et. al., "Radon concentrations in air and water in the thermal spas of Ischia Island." in Indoor and Built Environment, April 22, 2013.

Credits

Introduction Italy: A Land Born of the Sea

PAGE 1 *Satellite view of Italy*, http://www.nasa.gov/multimedia/imagegallery/image_feature_2349.html

PAGE 3 *Tectonic Plate Map of Italy,* © Ann C. Pizzorusso 2013, designed and illustrated by Matt Kania, Maphero Inc.

PAGE 4 *Scala dei Turchi,* SICILY, http://commons.wikimedia.org/wiki/File%3AScala_dei_Turchi.jpg

PAGE 5 *Italian Alps*, http://commons.wikimedia.org/wiki/File%3AItalian_Alps-Valley.jpg

PAGE 7 LEFT, *Satellite view of Sicily,* http://commons.wikimedia.org/wiki/File%3ASicily_with_Mount_Etna.jpg

PAGE 7 RIGHT, *Satellite view of Venice*, http://commons.wikimedia.org/wiki/File%3AVenice_iko_2001092.jpg

PAGE 8 *Mt. Etna,* http://commons.wikimedia.org/wiki/File%3AFranco_Zanghi_Etna_10.jpg

PAGE 9 *Locations mentioned in the Essays,* © Ann C. Pizzorusso 2013, designed and illustrated by Francesco Filippini

I. A Landscape Divine: Etruscans and their Environment

PAGE 11 *Pitigliano* © Ann C. Pizzorusso 2013

PAGE 12 *Etruscans*, antique print, author's collection

PAGE 13 LEFT, *Steaming landscape* © Ann C. Pizzorusso 2013

PAGE 13 RIGHT, *Columnar lava* © Ann C. Pizzorusso 2013

PAGE 15 *Geologic map of Etruria*, © Ann C. Pizzorusso 2013, designed and illustrated by Francesco Filippini

PAGE 16 Photo Christopher Wallston

PAGE 18 UPPER, *Bagnoregio* © Ann C. Pizzorusso 2013

PAGE 18 LOWER, *Caverns in tuff* © Ann C. Pizzorusso 2013

PAGE 19 *Weathered tuff* © Ann C. Pizzorusso 2013

PAGE 20 *Badia Bridge, Vulci*, http://commons.wikimedia.org/wiki/File:Ponte_Di_Badia_Vulci.jpg

PAGE 21 *Thermal waters*, http://commons.wikimedia.org/wiki/File:Cascate_del_Gorello_a_Saturnia.jpg

PAGE 22 LEFT, *Lake Bolsena,* © Ann C. Pizzorusso 2013

PAGE 22 RIGHT, *Lake Bolsena Island of Bisentina,* © Ann C. Pizzorusso 2013

PAGE 24 *Etruscan Cosmos*, © Ann C. Pizzorusso 2013, designed and illustrated by Francesco Filippini after Marziano Capello and Massimo Pallottino

PAGE 25 *Civita Castellana*, http://commons.wikimedia.org/wiki/File%3AJean-Baptiste_Camille_Corot_-_Civita_Castellana.jpg

PAGE 26 TOP, *Bagni San Filippo*, http://commons.wikimedia.org/wiki/File%3ABagniSanFilippoBalenaFossoBianco1.JPG

PAGE 26 TOP, *Cavernous Road*, Sovana,© Ann C. Pizzorusso 2013

PAGE 26 BOTTOM, *Cavernous Road*, Sovana,© Ann C. Pizzorusso 2013

PAGE 27 *Cavernous Road*, Sovana, http://upload.wikimedia.org/wikipedia/commons/7/7f/Via_Cava_Necropoli_Sovana.jpg

PAGE 28 *Etruscan priest*, bronze, by permission of the Vatican Museum Gregoriano Etrusco

PAGE 31 *Liver of Piacenza*, http://commons.wikimedia.org/wiki/File:Piacenza_Bronzeleber.jpg

PAGE 31 *Liver of Piacenza*, http://commons.wikimedia.org/wiki/File:Haruspex.png

PAGE 33 *Bird Augurer*, http://commons.wikimedia.org/wiki/File%3AA_dancing_man_from_the_painted_walls_of_the_tomb_of_the_Augurs_at_Tarquinia%2C_525-500_BCE%2C_Etruscan.jpg

PAGE 35 *Volcanic Lightning*, http://commons.wikimedia.org/wiki/File%3ARinjani_1994.jpg

PAGE 37 *Storm over Etruria*, © Ann C. Pizzorusso 2013, modified by Francesco Filippini

PAGE 39 *Etruscan priest*, NY Carlsberg Glyptotek, Copenhagen, photo Ann C. Pizzorusso

PAGE 40 *Aplu*, http://commons.wikimedia.org/wiki/File%3APopulonia_AR_10_asses_2340001.jpg

PAGE 41 top *Magnetite*, http://commons.wikimedia.org/wiki/File:Magnetite_Lodestone.jpg

PAGE 41 lower right, *Magnetite*, https://en.wikipedia.org/wiki/File:Magnetite.jpg

PAGE 41 lower left, *Fulgurite*, http://commons.wikimedia.org/wiki/File%3AFulgsdcr.jpg

PAGE 43 *Geomagnetic Map of Italy*, © Ann C. Pizzorusso 2013, adapted from the original published by Istituto Nazionale di Geofisica e Vulcanologia, designed and illustrated by Matt Kania, Maphero, Inc.

PAGE 45 *Volcanic map*, © Ann C. Pizzorusso 2013 designed and illustrated by Francesco Filippini

PAGE 47 *Etruscan landscape*, http://commons.wikimedia.org/wiki/File%3ALe_Balze_con_vista_sulla_Valdera.JPG

PAGE 48 *Antique Etruscan Architectural Print*, author's collection

PAGE 49 Cuniculi © Franco Ravelli

PAGE 50 *Antique Etruscan Architectural Print*, author's collection

PAGE 51 top, *Orvieto*, http://en.wikipedia.org/wiki/File:Orvieto_view.jpg

PAGE 51 bottom, *Orvieto*, http://commons.wikimedia.org/wiki/File:Joseph_Mallord_William_Turner_003.jpg

PAGE 52 *Antique Etruscan Architectural Print*, author's collection

PAGE 53 *Etruscan Tomb*, http://upload.wikimedia.org/wikipedia/

commons/thumb/f/fc/Tomba_dei_Rilievi_%28Banditaccia%29.jpg/800px-Tomba_dei_Rilievi_%28Banditaccia%29.jpg

PAGE 54 right, *Stone altars*, NY Carlsberg Glyptotek, Copenhagen, photo Ann. C. Pizzorusso

PAGE 54 left, *Stone altars*, NY Carlsberg Glyptotek, Copenhagen, photo Ann C. Pizzorusso

PAGE 55 *Tomb of the Leopards*, http://commons.wikimedia.org/wiki/File:Tarquinia_Tomb_of_the_Leopards.jpg

PAGE 56 *Blue Devil*, NY Carlsberg Glyptotek, Copenhagen, photo Ann C. Pizzorusso

PAGE 57 top, *Tomb of Orcus*, http://commons.wikimedia.org/wiki/File%3AAgamemnon_Tiresias_and_Ajax_Tomb_of_Orcus_II.jpg

PAGE 57 bottom, *Goddess of Underworld*, NY Carlsberg Glyptotek, Copenhagen, photo Ann C. Pizzorusso

PAGE 58 *Cinerary urn*, Altes Museum, Berlin, photo Ann C. Pizzorusso

II. Hell on Earth: into Virgil's Underworld

PAGE 59 *Aeneas and the Sibyl in the Underworld*, Brueghel the Elder, 1598 and rediscovered in 2001, Private Collection

PAGE 60 *Solfatara*, © Ann C.Pizzorusso 2013

PAGE 62 *Campi Flegrei*, http://upload.wikimedia.org/wikipedia/commons/f/ff/Naples_ali_2012191_lrg.jpg

PAGE 63 *Temple of Serapis* © Ann C. Pizzorusso 2013

PAGE 64 *Cuma* © Bonnie Alberts 2013

PAGE 65 *Temple of Apollo* © Bonnie Alberts 2013

PAGE 67 top, *Cuma, Cave of the Sibyl* © Ann C. Pizzorusso 2013, modified:Francesco Filippini

PAGE 67 bottom, left, *Cuma, Cave of the Sibyl* © Ann C. Pizzorusso 2013

PAGE 67 bottom, right, *Cuma, Cave of the Sibyl* © Bonnie Alberts 2013

PAGE 68 *Sibyl of Cuma*,
http://commons.wikimedia.org/wiki/File%3ACumaeanSibylByMichelangelo.jpg

PAGE 70 left, *Lake Averno*, antique etching, author's collection

PAGE 70 right, *Lake Averno*, © Bonnie Alberts 2013

PAGE 71 *Grotto of the Dog*, antique etching, author's collection

PAGE 72 *Map of the Underworld*, © Ann C. Pizzorusso 2013, designed and illustrated by Francesco Filippini

PAGE 73 *Grotto of the Sibyl*, antique etching, author's collection

PAGE 74 *Lake Lucrino*, © Bonnie Alberts 2013

PAGE 75 *Cavern* © Bonnie Alberts 2013

PAGE 77 top, *Solfatara* © Ann C. Pizzorusso 2013 modified by Francesco Filippini

PAGE 77 bottom left, *Solfatara* © Ann C. Pizzorusso 2013

PAGE 77 bottom right, *Campi Flegrei*, Italian Civil Defense Department

PAGE 78 *Balneum Tripergulae*, http://commons.wikimedia.org/wiki/File%3APetrus_de_Ebulo_-_Balneum_Tripergulae.jpg

PAGE 79 top left, *De Balneis Puteolanis*, http://commons.wikimedia.org/wiki/File%3APetrus_de_Ebulo_-_Balneum_Sudatorium.jpg

PAGE 79 top right, *Balneum Sulphatara*, http://commons.wikimedia.org/wiki/File%3APetrus_de_Ebulo_-_Balneum_Sulphatara.jpg

PAGE 79 bottom left, *Balneum Bullae*, http://commons.wikimedia.org/wiki/File%3APetrus_de_Ebulo_-_Balneum_Bullae.jpg

PAGE 79 bottom right, *Balneum Spelunca*, http://commons.wikimedia.org/wiki/File%3APetrus_de_Ebulo_-_Balneum_Spelunca.jpg

PAGE 80 *Vesuvius*, http://eoimages.gsfc.nasa.gov/images/imagerecords/1000/1045/aster_vesuvius_lrg.jpg

PAGE 81 *Goethe and the Gulf of Naples*, Heinrich Christoph Kolbe, 1826, in the collection of the Friedrich Schiller University, Jena, Inv. Number GP 276, (Photo Peter Scheere)

PAGE 82 left, *Mt. Etna from Taormina*, Thomas Cole, 1843, http://commons.wikimedia.org/wiki/File%3ACole_Thomas_Mount_Aetna_from_Taormina_1844.jpg

PAGE 82 right, *Mt.Etna*, http://upload.wikimedia.org/wikipedia/commons/3/31/Etna_eruption_seen_from_the_International_Space_Station.jpg

PAGE 83 *Vesuvius* © Bonnie Alberts 2013

PAGE 85 *Baia underwater*, by permission of the Minister of Cultural Affairs—under the direction of the Superintendent of Archeological Affairs of Naples and Pompeii

PAGE 86 *Baia underwater*, by permission of the Minister of Cultural Affairs—under the direction of the Superintendent of Archeological Affairs of Naples and Pompeii

PAGE 87 *Baia underwater*, by permission of the Minister of Cultural Affairs—under the direction of the Superintendent of Archeological Affairs of Naples and Pompeii

PAGE 88 *Baia, Thermal Baths* © Bonnie Alberts 2013

PAGE 89 *Baia, Thermal Baths* © Bonnie Alberts 2013

III. Paradise Bejeweled

PAGE 91 Doré, Gustav, *Illustrations for Dante's Divine Comedy*, Purgatory, Canto XXX, http://www.wikipaintings.org/en/gustave-dore/purgatorio-canto-30

PAGE 92 *Antique mineral print*, author's collection

PAGE 93 *Dante and the Divine Comedy,* Domenico di Michelino, 1465, Santa Maria del Fiore, fresco, Florence, http://commons.wikimedia.org/wiki/File:Dante_Domenico_di_Michelino_Duomo_Florence.jpg

PAGE 95 *Brooch*, 1335, Venice, Museum of Castelvecchio, Verona, Photo: Umberto Tomba

PAGE 97 *Gems,* © Ann C. Pizzorusso 2013, designed and illustrated by Francesco Filippini

PAGE 98 *Caterina de'Medici*, Uffizi Gallery, Florence, http://en.wikipedia.org/wiki/File:KatharinavonMedici.jpg

PAGE 103 *Pearl,* © Ann C. Pizzorusso 2013, designed and illustrated by Francesco Filippini

PAGE 104 *Pearls*, http://commons.wikimedia.org/wiki/File%3AShell_and_pearls.JPG

PAGE 113 *Asterism*, © Ann C. Pizzorusso 2013 designed and illustrated by Francesco Filippini

PAGE 115 *Star Sapphire*, American Museum of History, New York, http://en.wikipedia.org/wiki/File:Star_of_India_Gem.JPG

PAGE 118 *Allegory of the Catholic Faith*, Vermeer, Johannes, Metropolitan Museum of Art, New York, http://commons.wikimedia.org/wiki/File:Vermeer_The_Allegory_of_the_Faith.jpg

PAGE 123 Doré, Gustav, *Illustrations for Dante's Divine Comedy*, Paradise,

Canto xiv, http://it.wikipedia.org/wiki/File:Par_14_dore.jpg

PAGE 124 *Emerald* © Ann C. Pizzorusso 2013, designed and illustrated by Francesco Filippini

PAGE 125 Photo: John Mirabella

PAGE 127 Doré, Gustav, *Illustrations for Dante's Divine Comedy*, Paradise, Canto XIV, http://it.wikipedia.org/wiki/File:Par_27.jpg

PAGE 128 *Solar System*, http://en.wikipedia.org/wiki/File:Solar_System_size_to_scale.svg

PAGE 129 *Dante's Cosmology*, http://commons.wikimedia.org/wiki/File%3APtolemaicsystem-small.png

PAGE 130 Doré, Gustav, *Illustrations for Dante's Divine Comedy*, Paradise, Canto xiii http://it.wikipedia.org/wiki/Paradiso_-_Canto_tredicesimo

PAGE 132 *Zodiac woodcut*, XVI century, http://en.wikipedia.org/wiki/File:Zodiac_woodcut.png

PAGE 133 Marbode, *De Gemmarum Lapidum*, bookplate, 1539, antique print, author's collection

PAGE 135 *Mineral print*, author's collection

PAGE 137 *Stairway to Heaven*, Raphael, fresco, Vatican, © Ann Jones 2013

PAGE 138 *Amber*, http://en.wikipedia.org/wiki/File:Gouttes-drops-resine-2.jpg

PAGE 139 *Relief of a Woman and Youth Reclining*, late VI-early V century B.C., Etruscan, amber, Gift of J. Pierpont Morgan, Metropolitan Museum of Art, New York

PAGE 141 *Amber*, http://commons.wikimedia.org/wiki/File%3AInsects_in_baltic_amber.jpg

PAGE 142 *Antique Mineral Print*, author's collection

IV. Leonardo da Vinci: On the Nature of Things

PAGE 143 *Sheet of water studies and notes on atmosphere,* Leonardo da Vinci, Pen and ink and red chalk, (RL 12661) Royal Collection Trust/© Her Majesty Queen Elizabeth II 2013

PAGE 145 top left, *Fish fossil,* (*Coelodus costai)*, Natural History Museum, Milan, Italy, http://commons.wikimedia.org/wiki/File:9178_-_Milano_-_Museo_storia_naturale_-_Coelodus_costai_-_Foto_Giovanni_Dall%27Orto_22-Apr-2007.jpg

PAGE 145 top right, *Arietites fossil,* http://commons.wikimedia.org/wiki/File%3AArietenpflaster_von_Boldelshausen.JPG

PAGE 145 bottom, *Fern fossil,* http://commons.wikimedia.org/wiki/File%3ACycadopteris_jurensis.JPG

PAGE 147 *Stream running through a rocky ravine, with birds in the foreground,* Leonardo da Vinci, Pen and ink (RL 12395) Royal Collection Trust/© Her Majesty Queen Elizabeth II 2013

PAGE 148 *Sheet of water studies, with notes,* Leonardo da Vinci, Pen and ink and red chalk on coarse yellowish paper (RL 12662) Royal Collection Trust/© Her Majesty Queen Elizabeth II 2013

PAGE 149 *Bird's-eye view of a river valley with canal,* Leonardo da Vinci, Pen and ink over black chalk (RL 12398) Royal Collection Trust/© Her Majesty Queen Elizabeth II 2013

PAGE 151 *Sheet of notes on how to represent the Deluge, with marginal illustrations,* Leonardo da Vinci, Pen and ink, (RL 12660v) Royal Collection Trust/© Her Majesty Queen Elizabeth II 2013

PAGE 152 *Sheet of notes on how to represent the Deluge, with marginal illustrations,* Leonardo da Vinci, Pen and ink, (RL 12665r) Royal Collection Trust/© Her Majesty Queen Elizabeth II 2013

PAGE 154 *Watersheds of the Arno,* Leonardo da Vinci (RL 12277) Royal Collection Trust/© Her Majesty Queen Elizabeth II 2013

PAGE 155 *Town at center of vortex,* Leonardo da Vinci, Black chalk on coarse paper (RL 12378) Royal Collection Trust/© Her Majesty Queen Elizabeth II 2013

PAGE 156 *Water breaking through a mountain pass, causing a rockslide and large waves in a lake,* Leonardo da Vinci, Pen and two inks (black and yellow) over black chalk, with wash on coarse brownish paper (RL 12380) Royal Collection Trust/© Her Majesty Queen Elizabeth II 2013

PAGE 157 *Vortex flow pattern,* http://commons.wikimedia.org/wiki/File:TC_Crystal_27_dec_2002_0630Z.jpg

PAGE 158 *Natural umber powdered,* http://commons.wikimedia.org/wiki/File%3ATerra_ombra_naturale_umber.jpg

PAGE 159 *Lazurite,* http://commons.wikimedia.org/wiki/File:Afghanite_et_lazurite_sous_UV_%28Afghanistan%29_.JPG

PAGE 160 *Azurite and Malachite,* http://commons.wikimedia.org/wiki/File:Azurite-Malachite_from_Marocco.jpg

PAGE 161 left, *Orpiment,* photo Ann C. Pizzorusso

PAGE 161 right, *Realgar,* http://commons.wikimedia.org/wiki/File:Realgar,_1Rumunia1,_Baia_Sprie.jpg

PAGE 162 *Stucco wall fragment,* Villa of Tiberius, Sperlonga, Italy, I century A.D., photo Ann C. Pizzorusso

V. Leonardo's Geology: A Tale of Two Paintings

PAGE 163, left, *Virgin of the Rocks*, Louvre, Paris, ArtResource

PAGE 163, right, *Virgin of the Rocks*, National Gallery, London, ArtResource

PAGE 165 *Virgin of the Rocks*, Louvre, Paris, detail, ArtResource

PAGE 166 *Virgin of the Rocks*, Louvre, Paris, ArtResource, modified with geological captions © Ann C. Pizzorusso

PAGE 167*Virgin of the Rocks*, National Gallery, London, ArtResource, modified with geological captions © Ann C. Pizzorusso

PAGE 168 *Star of Bethlehem, wood anemone and sun spurge,* Leonardo da Vinci, 1505-10, The Royal Collection © Her Majesty Queen Elizabeth II, 2013

PAGE 169 detail, upper, *Virgin of the Rocks*, Louvre, Paris, ArtResource

PAGE 169 detail, lower, *Virgin of the Rocks*, National Gallery, ArtResource

PAGE 170 top left, *Virgin and St. Anne*, Louvre, Paris, ArtResource

PAGE 170 center right, *Virgin and St. Anne*, Louvre, Paris, detail, ArtResource

PAGE 170 bottom, *An outcrop of Stratified Rock,* Leonardo da Vinci, c. 1510, (RL 12394) Royal Collection Trust © Her Majesty Queen Elizabeth II, 2013

PAGE 171 *Virgin of the Rocks*, Louvre, Paris, detail, ArtResource

PAGE 173 *Angel*, Ambrogio de Predis, c. 1506, Museum of the Church of Santa Maria delle Grazie, Milan, http://commons.wikimedia.org/wiki/File:Ambrogio_de_Predis_010.jpg

PAGE 174 Altar from San Francesco il Grande, Milan, antique print, author's collection

PAGE 176 *Annunciation*, Leonardo da Vinci, c. 1472, http://commons.wikimedia.org/wiki/File:Leonardo_da_Vinci_-_Annunciazione_-_Google_Art_Project.jpg

PAGE 177 *Burlington Cartoon*, Leonardo da Vinci, 1499, National Gallery http://commons.wikimedia.org/wiki/File:Leonardo_-_St._Anne_cartoon-alternative-downsampled.jpg

PAGE 179 *The Resurrection of Christ with Saints Leonardo of Noblac and Lucia*, Giovanni Boltraffio and Marco d'Oggiono, 1492, Gemaldegalerie der Staatliche Museen, Berlin

IV. Recesses of the Mind and Soul

PAGE 181 *Grotto of Castellana*, http://en.wikipedia.org/wiki/File:Grotte_Castellana_03apr06_04.jpg

PAGE 182 *Catacombs of Naples*, antique etching, author's collection

PAGE 184 Courtesy of Grotte di Frasassi

PAGE 185 left, *Limestone column,* http://commons.wikimedia.org/wiki/File%3AGrotte_du_Grand_Roc__Stalactite_and_stalagmite_-_20090923.jpg

PAGE 185 center, *Limestone flowstone,* http://commons.wikimedia.org/wiki/File:flowstone_%22veils%22_climb_nearly_40_feet_towards_the_ceiling_in_bridal_cave_near_camdenton._hundreds_of_caves_honeycomb..._-_nara_-_551358.tif

PAGE 185 *Limestone soda straws,* http://en.wikipedia.org/wiki/File:Gardeners_Guts_Speleothem_Straws.jpg

PAGE 186 *Limestone Cavern Formation*, © Ann C. Pizzorusso 2013 designed and illustrated by Francesco Filippini

PAGE 189 left, *Neapolitan Water Merchant*, antique etching, author's collection

PAGE 189 right, *Etruscan statues*, Etruscan Museum Villa Giulia, by permission of the Superintendent of Archeological Affairs of Southern Etruria

PAGE 190 left, *Grotto of Scaloria*, by permission of the Minister of Cultural Affairs, Superintendent of Archeological Affairs of Pulgia

PAGE 190 right, *Stalagmite from Grotto of Scaloria*, by permission of the Museum of Archeology Bari and the Superintendent of Archeological Affairs of Pulgia photo © Bonnie Alberts 2013

PAGE 191 *Stalactite drip water*, photo Thomas Bresson http://commons.wikimedia.org/wiki/File%3AThomas_Bresson_-_Stalactite-5_(by).jpg

PAGE 191 *Stalactite drip water*,

http://my.opera.com/orsoyoghurt/albums/showpic.dml?album=449970&picture=9747708

PAGE 192 *Capitoline Wolf*, © Ann C. Pizzorusso 2013

PAGE 193 left, *Votive offerings*, Etruscan Museum Villa Giulia, by permission of the Superintendent of Archeological Affairs of Southern Etruria

PAGE 193, right, *Wax votive candles* © Bonnie Alberts 2013

PAGE 194 left, *Grotto of San Michele* © Bonnie Alberts 2013

PAGE 194 right, *Grotto of San Michele*, © Ann Pizzorusso 2013

PAGE 195 *St. Francis Receiving the Stigmata,* Giotto, http://commons.wikimedia.org/wiki/File:Giotto_-_Sankt_Franciskus_stigmatisering.jpg

PAGE 197 *Grotto*, Villa Castello, Florence, http://commons.wikimedia.org/wiki/File%3AVilla_Castello_Florence_Apr_2008.jpg

PAGE 199 *Grotto*, Boboli Gardens, http://commons.

wikimedia.org/wiki/File:Boboligrotto.jpg

PAGE 200 *Grotto of Tiberius*, Sperlonga © Ann C. Pizzorusso 2013

PAGE 201 *Grotto*, Boboli Gardens, Florence http://commons.wikimedia.org/wiki/File:Grotta_del_buontalenti,_esterno_dettaglio_01.jpg

PAGE 203 *Fountain of Diana*, Tivoli, Villa 'Este, Rome, © Ann C. Pizzorusso 2013

PAGE 206 *Votive bas relief*, Archeological Museum of Naples, by permission of the Minister of Cultural Affairs and the Superintendent of Archeological Heritage for Naples and Pompeii

PAGE 207 *Maps of the thermal waters of Ischia* adapted from Terme di Ischia and Geothermalenergy.it, © Ann C. Pizzorusso 2013, designed and illustrated by Matt Kania, Maphero,Inc.

PAGE 208 *Votive bas relief*, Archeological Museum of Naples, by permission of the Minister of Cultural Affairs and the Superintendent of Archeological Heritage for Naples and Pompeii

PAGE 209 *Votive bas relief*, Archeological Museum of Naples, by permission of the Minister of Cultural Affairs and the Superintendent of Archeological Heritage for Naples and Pompeii

PAGE 210 *Ischia* © Ann C. Pizzorusso 2013

PAGE 211 *Pale di San Martino*, http://commons.wikimedia.org/wiki/File%3APale_di_san_martino_tramonto.jpg

About the Author

Ann Pizzorusso is a geologist and Italian Renaissance scholar. After many years of doing virtually everything in the world of geology— drilling for oil, hunting for gems, cleaning up pollution in soil and groundwater, she turned her geologic skills toward Leonardo da Vinci.